FIXING
MY
GAZE

FIXING
MY
GAZE

A SCIENTIST'S JOURNEY
INTO SEEING IN
THREE DIMENSIONS

SUSAN R. BARRY

BASIC
BOOKS

A Member of the Perseus Books Group
New York

Grateful acknowledgment is given for permission to reprint the previously published material: Excerpt from "Two Tramps in Mud Time" from *The Poetry of Robert Frost*, edited by Edward Connery Lathem. Copyright 1969 by Henry Holt and Company. Copyright 1936 by Robert Frost, copyright 1964 by Lesley Frost Ballantine. Reprinted by permission of Henry Holt and Company, LLC; excerpt from J657/F466 "I dwell in Possibility." Reprinted by permission of the publishers and the Trustees of Amherst College from *The Poems of Emily Dickinson*, Thomas H. Johnson, ed., Cambridge, Mass.: The Belknap Press of Harvard University Press, Copyright © 1951, 1955, 1979, 1983 by the President and Fellows of Harvard College.

Note: The information in this book is true and complete to the best of our knowledge. This book is not a substitute for diagnosis and treatment by a licensed vision care professional. In no way is this book intended to replace, countermand, or conflict with the advice given to you by your own vision care professional. The ultimate decision concerning care should be made between you and your doctor. We strongly recommend you follow his or her advice. Information in this book is general and is offered with no guarantees on the part of the author or Basic Books. The authors and publisher disclaim all liability in connection with the use of this book.

Designed by Jeff Williams

Library of Congress Cataloging-in-Publication Data
Barry, Susan R.
    Fixing my gaze : a scientist's journey into seeing in three dimensions / Susan R. Barry.
        p. cm.
    Includes bibliographical references and index.
    ISBN 978-0-465-00913-8 (alk. paper)
    1. Barry, Susan R.—Health. 2. Strabismus—Patients—United States—Biography. 3. Depth perception. 4. Behavioral optometry. 5. Visual training. 6. Neurobiologists—United States—Biography. I. Title.

RE771.B37 2009
617.7'62—dc22

                                                        2009008900

10 9 8 7 6 5 4 3

In memory of my mother,
Estelle Florence Fisher Feinstein,
a woman who saw
in great depth.

Contents

Note to the Reader   ix
Foreword by Oliver Sacks   xi

1   Stereoblind   1
2   Mixed-Up Beginnings   17
3   School Crossings   35
4   Knowing Where to Look   47
5   Fixing My Gaze   69
6   The Space Between   89
7   When Two Eyes See as One   105
8   Nature and Nurture   133
9   Vision and Revision   155

Acknowledgments   167
Glossary   171
Resources   175
Notes   179
Index   237

Note to the Reader

All of the stories in this book are true, but the names of some of the people have been changed to respect their privacy.

Foreword
by Oliver Sacks

I first met Sue Barry in 1996 at a launch party for her husband, Dan, an astronaut. We soon got to talking about different ways of experiencing the world—how Dan, for example, in the microgravity of spaceflight, had no direct sense of up or down, so he had to find other ways of orienting himself in space. She herself, Sue said, experienced the world in an unusual way as a consequence of having developed crossed eyes, or strabismus, in early infancy. Her eyes had been straightened by surgery, and she had 20/20 vision in both eyes, but they were still not fully aligned. Her brain had learned to suppress the image from one eye or the other so that she did not experience a confusing double vision. Normally the brain constructs a perception of depth by comparing the images from the two eyes, but in Sue's case, where one or the other image was suppressed, no such comparison was possible. So, though she had learned to judge distance and depth by other cues, she had never experienced true "solid vision," or stereoscopy. Her world was entirely flat.

But all in all, she said, she got along perfectly well—she drove a car, she could play softball, she could do whatever anyone else could. She might not be able to see depth directly, as other people could, but she could judge it as well as anybody, using other cues, such as perspective, occlusion, shading, or motion parallax. It was no big deal.

I asked Sue if she could imagine what the world would look like if viewed stereoscopically, and she said, yes, she thought she could—after all, she was a professor of neurobiology, and she had read plenty of papers on visual processing, binocular vision, and stereopsis. She felt this knowledge had given her some special insight into what she was missing—she knew what stereopsis must be like, even if she had never experienced it.

But in December 2004, almost nine years after our initial conversation, she wrote me a letter, which began, "You asked me if I could imagine what the world would look like when viewed with two eyes. I told you that I thought I could. . . . But I was wrong."

She could say this with some conviction because she had suddenly, unexpectedly, acquired stereovision herself, and the reality of this, the actual experience, was utterly beyond anything imagination could have conceived. She was almost fifty and had been having increasing visual difficulties resulting from the misalignment of her eyes. Finally, she had embarked on an intensive course of training with a developmental optometrist, and one day, after learning to align her eyes, she suddenly saw the steering wheel of her car "popping out" from the dashboard. After having lived in a flat world for fifty years, Sue felt this sudden leap into three-dimensionality as a revelation. Her world was now full of a new sort of visual beauty and wonder so deep that three years later, when she wrote to me, she was still enraptured with it.

"My new vision continues to surprise and delight me," she wrote.

One winter day, I was racing from the classroom to the deli for a quick lunch. After taking only a few steps from the classroom building, I stopped short. The snow was falling lazily around me in large, wet flakes. I could see the space between each flake, and all the flakes together produced a beautiful three-dimensional dance. In the past, the snow would have appeared to fall in a flat sheet in one plane slightly in front of me. I would have felt like I was looking in on the snowfall. But now, I felt myself within the snowfall, among the snowflakes. Lunch forgotten, I watched the snow fall for several minutes, and, as I watched, I was overcome with a deep sense of joy. A snowfall can be quite beautiful—especially when you see it for the first time.

Most of the phone calls and letters I receive are about mishaps, problems, losses of various sorts. Sue's letter, though, was a story not of loss and lamentation but of the sudden gaining of a new sense and sensibility and, with this, a sense of delight and jubilation. Yet, her letter also sounded a note of bewilderment and reservation: she did not know of any experience or story like her own and was perplexed to find, in all she had read, that the achievement of stereoscopy in adult life was "impossible."

Indeed, what Sue described to me in her letter went completely against the current dogma of "critical periods" in sensory development—the notion that stereoscopy (like many other aspects of visual perception, and like language, as well)—had to be acquired in the first three or four years of life, or it could never be acquired, for the critical brain cells and circuitry needed for stereovision would fail to develop.

Long suspected by surgeons operating on children with strabismus, this notion of a critical period seemed to be confirmed by the famous experimental work of David Hubel and Torsten Wiesel, who showed that if kittens were rendered strabismic by detaching an eye muscle, binocular depth cells would fail to develop in their brains, and they would lack stereovision. It was only when Sue learned of these experiments—she was a college student at the time—that she realized she herself might be stereoblind, like the kittens. This, indeed, is the vivid opening scene in her narrative:

Stereoblind? Was I stereoblind? I looked around. The classroom didn't seem entirely flat to me. I knew that the student sitting in front of me was located between me and the blackboard because the student blocked my view of the blackboard. When I looked outside the classroom window, I knew which trees were located further away because they looked smaller than the closer ones. The footpath outside the window appeared to narrow as it extended out into the distance. Through cues like these, I could judge depth and distance. I knew the world was in 3D. Yet, my professor implied that there was another, different way to see space and depth. He called this way of seeing stereopsis. I couldn't imagine what he was talking about.

When Sue next went to her eye doctor for a routine check, she asked him to check whether she had stereovision. He brought out a stereoscope and test stereo pictures. Sue could not "get" any of them, could not imagine what "getting" them would be like. Would it be possible for her to gain stereovision, she asked? The doctor replied, no, it was much too late, and added, "Stereopsis is just a little fine-tuning for the visual system. You don't need stereovision because you don't have stereovision."

Sue accepted that she would never have stereovision and got on with her life, becoming a teacher and researcher, marrying, and raising a family. Somewhere, at the back of her mind—for she is a scientist and incessantly curious about how the world works—was a question: what could stereovision be like? And yet, her life, visually and otherwise, was full and rich, and she did not "miss" stereo or think of it too much. So thirty years later, when she finally sought vision therapy and unexpectedly gained stereovision, this came as a bonus, a miraculous complement to her other visual improvements.

Sue exulted in her newfound sense of stereoscopic depth. She found it much more than "fine-tuning"—it was an entirely new way of seeing. "People who have always had stereopsis," she said to me, "take it for granted. They have no idea how wonderful it is. You have to have been stereoblind for half a century and then acquire it to value it properly."

How was Sue able to acquire, essentially, a whole new sense so long after the "critical period"? I was as puzzled by Sue's story as she was. And I was intrigued, for I myself had never taken stereoscopy for granted. On the contrary, I was something of a stereophile, having played with 3D drawings and Victorian stereo viewers as a child and later experimented with stereo photography. So I arranged to meet Sue again, and this inspired me to write an article about her experiences in 2006.*

But that was not the end of the matter.

All that Sue intimated to me in her letters and conversations has now been expanded and deepened into a fascinating account. *Fixing My Gaze* is a beautiful description and appreciation of two very distinct ways of seeing—with and without the benefit of stereoscopy. But it is also an exploration of much more. Sue is at pains not only to present her story in clear, lucid, often poetic language, but also, as a scientist, to provide explanation and understanding.

She is in a unique position to do this, drawing on both her personal experience and her background as a neurobiologist. She has interviewed many eminent vision researchers and pondered the problem of critical periods with them. Her experience indicates that there seems to be sufficient plasticity in the adult brain for these binocular cells and circuits, if some have survived the critical period, to be reactivated later. In such a situation, though a person may have had little or no stereovision that she can remember, the potential for stereopsis is nonetheless present and may spring to life—most unexpectedly—if good alignment of the eyes can

*"Stereo Sue," *The New Yorker,* June 19, 2006, 64–73.

be obtained. That this seems to have happened with Sue after a dormant period of almost fifty years is very striking.

Although Sue originally thought her own case unique, she has since found a number of other people with strabismus and related problems who have unexpectedly achieved stereovision through vision therapy. This is no easy accomplishment. It may require not only optical corrections (proper lenses or prisms and so forth) but very intensive training and learning—in effect, one must learn how to align the eyes and fuse their images, while unlearning the unconscious habit of suppressing vision, which has been occurring perhaps for decades. In this way, vision therapy is directed at the whole person: it requires high motivation and self-awareness, as well as enormous perseverance, practice, and determination, as does psychotherapy, for instance, or learning to play the piano. But it is also highly rewarding, as Sue brings out. And this ability to acquire new perceptual abilities later in life has great implications for anyone interested in neuroscience or rehabilitation and, of course, for the millions of people who, like Sue, have been strabismic since infancy. Sue's case, together with many others, suggests that if there are even small islands of function in the visual cortex, there may be a fair chance of reactivating and expanding them in later life, even after a lapse of decades, if vision can be made optically possible. Cases like these may offer new hope for those once considered incorrigibly stereoblind. *Fixing My Gaze* will offer inspiration for anyone in this situation, but it is equally a very remarkable exploration of the brain's ability to change and adapt, as well as an ode to the fascination and wonder of the visual world, even those parts of it which many of us take for granted.

1

STEREOBLIND

But yield who will to their separation,
My object in living is to unite
My avocation and my vocation
As my two eyes make one in sight.

—"Two Tramps in Mud Time," by Robert Frost

I was twenty years old and a college student before I learned that I did not see the way other people did. This surprising news came to me as I listened to a lecture on vision in my college neurobiology class. On that gray November morning, I felt sleepy and sluggish, but something my professor said jolted me out of my inattentive state. He was describing the development of the visual system, highlighting experiments done on walleyed and cross-eyed kittens. Cats, like people, have two forward-facing eyes that they move together in coordinated ways. But the kittens in these studies had strabismus, or misaligned eyes. My professor mentioned that vision in these kittens had not developed normally. They probably couldn't see in 3D. In fact, many scientists and doctors assumed that the cats would never acquire stereovision, even if their eyes were later straightened, because this ability could develop only during a "critical period" in early life. What was thought to be true for cats was also believed to be true for people.

I was floored. My eyes had crossed when I was about three months old. When I looked at an object with my left eye, my right eye turned in, and when I looked with my right eye, my left eye moved noseward. But I had three eye-muscle surgeries at ages two, three, and seven, and these operations had aligned my eyes so that my eyes looked normal almost all the time. Surely, I saw normally too. Throughout childhood, I had 20/20 acuity with each eye and assumed that I had perfect vision.

Yet, I had just learned that people like me were missing a fundamental way of seeing. Fully alert now, I listened carefully to the professor's explanation. We have two eyes, he said, but only one view of the world. Since our eyes are separated on our face by our nose, they see from a slightly different perspective. It is in the brain that the images from the two eyes merge into one. For most people, this happens effortlessly. Both eyes are aimed at the same point in space, and the information from each is combined in the brain. The result is a sharply outlined, detailed, and depth-filled view of the world.

My professor added that a strabismic (or person with misaligned eyes) is not so lucky. Since a strabismic's eyes are not aimed at the same point in space, the difference between the left- and right-eye views is too great for the brain to combine the images into a single picture. The strabismic is confronted with a serious perceptual problem: she must somehow create a single, coherent worldview from conflicting input from the two eyes. To solve this problem, many strabismics suppress the information from one eye and look through the other. Some always use the same eye, while others continually switch between the two eyes, but in either case, they may never see normally through the two eyes together. As a result, most strabismics have reduced or absent stereovision. The professor concluded the lecture by saying that many strabismics don't see in 3D. They're virtually stereoblind.

Stereoblind? Was I stereoblind? I looked around. The classroom didn't seem entirely flat to me. I knew that the student sitting in front of me was located between me and the blackboard because the student blocked my view of the blackboard. When I looked outside the classroom window, I knew which trees were located further away because they looked smaller than the closer ones. The footpath outside the window appeared to narrow as it extended out into the distance. Through cues like these, I could judge depth and distance. I knew the world was in 3D. Yet, my professor implied that there was another, different way to see space and depth. He called this way of seeing stereopsis. I couldn't imagine what he was talking about.

After the lecture, I went directly to the college library and struggled through the scientific papers on vision. I spent the rest of the semester studying the subject and wrote my term paper on changes to the visual system of cats that started out life with misaligned eyes. I learned that the brain processes vision in a region in the back of the cerebral cortex called the visual cortex. Neurons from the retina in the back of the eye communicate over several synaptic connections with neurons in the visual cortex, and these cortical neurons are either "monocular" or "binocular." I learned that monocular neurons respond with nerve impulses to light stimuli coming from only the right or left eye, while binocular neurons respond to input from either eye. The majority of neurons in the visual cortex are binocular. However, strabismic infants have neurons that respond to the right or the left eye, but not both. The loss of binocular neurons results in a loss of normal binocular vision and stereopsis.

As I stayed up late reading through all these papers, I realized that I too might have a "monocular brain." Most of the neurons in my visual cortex probably responded to input from either my right or left eye, but not both. Although I no longer looked grossly cross-eyed as I had as a child, my eyes still wandered out of alignment on occasion, especially when I was tired. So, I always avoided looking people directly in the eye. Now I suspected that I was not only a little cross-eyed but also stereoblind.

On my next trip to the eye doctor for a routine eye exam, I asked about stereovision. The doctor was surprised by my concern and interest but got out his stereo tests. I flunked them all. He shrugged his shoulders and explained that I did not fuse the images provided by my two eyes. I saw the input from only one eye at a time and switched rapidly between them.

"Don't worry," he told me. "Stereopsis is just a little fine-tuning for the visual system." Then, he added, "You don't need stereovision because you don't have stereovision," a statement whose logic escapes me to this day.

Was stereopsis just a little fine-tuning for the visual system, or was it an important component of everyday seeing? Indeed, the answer to this question has eluded scientists for years. In fact, the entire phenomenon of stereopsis escaped scientists for centuries. Many of the great students of optics, including Euclid, Archimedes, da Vinci, Newton, and Goethe, never figured out how we see in stereoscopic depth. This role fell to a brilliant, but reticent, inventor by the name of Charles Wheatstone.

Wheatstone, a British scientist working in the early and mid-1800s, became the first person to measure the speed of electricity and was instrumental in the development of the first telegraph. He also discovered that the difference in viewing perspective between our two eyes is not an imperfection in our vision. Instead, this dif-

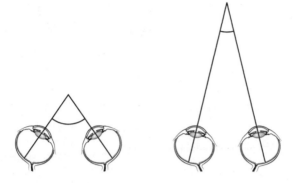

FIGURE 1.1: Your eyes turn in, or converge, to fixate, or look directly at, a near object; they turn out, or diverge, to fixate a more distant target. The straight lines in the figure indicate the lines of sight for each eye. (© Margaret C. Nelson)

ference provides us with stereopsis, or a depth-filled way of seeing the world.

Wheatstone knew that we turn in our eyes to look at nearby objects and turn them out to look at targets further away (Figure 1.1). You can determine this for yourself by asking a friend to follow the tip of an upright pencil held straight in front of him. As the pencil is brought closer to his face, he will turn in (converge) his eyes. As the pencil moves away, his eyes will turn out (diverge). These vergence movements cause the image of the pencil to fall on corresponding points of the two retinas where the light-sensing cells are found.

In order for you to see an object, light rays bouncing off it enter your eye and travel to the back of the eyeball where they land on the retina (Figure 1.2). In the retina, the rod and cone cells sense the light and pass this information on to other retinal cells and ultimately to neurons deeper in your brain.

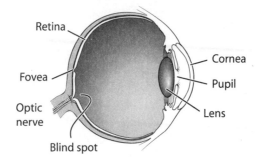

Retina

Fovea

Optic
nerve

Blind spot

Cornea

Pupil

Lens

FIGURE 1.2: The human eye including the pupil, lens, and retina. The central region of the retina is called the macula, and the center of the macula is the fovea. (© Margaret C. Nelson)

We can divide each retina into three regions: the fovea, or central region, the right side, and the left side. When you look directly at an object, its image falls on corresponding points on the central (foveal) region of both retinas. Other objects that cast their images on regions that are the same distance and in the same direction from each fovea also project to corresponding retinal points. Imagine that you are looking directly at the toy block in figure 1.3. The teddy bear located to your left casts its image on corresponding points on the right side of both your retinas, while the rattle, to the right, casts its image on corresponding points on the left side of both retinas.

In 1838, Wheatstone explained how the relative position of the images on the two retinas allows us to see in 3D. He published a paper quaintly titled "Contributions to the Physiology of Vision.—Part the First. On some remarkable, and hitherto unobserved, Phenomena of Binocular Vision." He explained how stereopsis works by introducing a model of the first stereoscope, so named because "stereo" is the Greek word for "solid," and images seen through the stereoscope look solid and real.

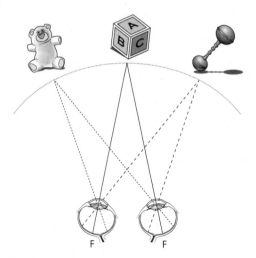

FIGURE 1.3: F refers to fovea. (© Margaret C. Nelson)

In the drawing of a stereoscope in Figure 1.4 on page 8, the two A's in the center represent mirrors oriented at 90° to each other. To use the stereoscope, you place your nose right at the juncture between the two mirrors. In this way, the right eye can see only the reflected version of a photograph placed at E in the figure, while the left eye can see only the reflected version of the image at E on the other side. Wheatstone placed into right slot E a mirror-image picture of an object as it would be seen by your right eye and into left slot E a mirror-image picture of the same object as it would be seen by your left eye. If you were to look into the stereoscope, your brain would fuse the two images into one, and you'd see the image in stereoscopic depth.

Wheatstone highlighted several figures to be used in his stereoscope, including the one illustrated in figure 1.5 on page 9. Each member of the pair shows a small square surrounded by a larger one. The figures look flat. If you were to cut out these two drawings and put one on top of the other, the large squares would

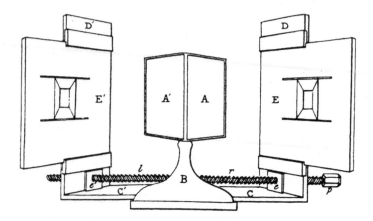

FIGURE 1.4: Wheatstone's illustration of his stereoscope. (Wheatstone C. 1838. Contributions to the Physiology of Vision.—Part the First. On some remarkable, and hitherto unobserved, Phenomena of Binocular Vision. *Philosophical Transactions of the Royal Society of London* 128: 371–94)

overlap perfectly but the small squares would not. Now, imagine that the figures were placed into the E slots in the stereoscope. When you looked into the stereoscope, your left eye would see only the reflected image of figure A, the left-hand figure, while your right eye would see only the reflected image of figure B. If you can see in 3D, your brain will fuse figures A and B, causing you to see just one small and one large square. As you look into the stereoscope, the edges of the larger, outer squares will fall on corresponding points of your two retinas, while the edges of the smaller, inner squares will not. This difference will cause the fused image of the large and small square to appear in different depth planes.

Using the stereoscope, Wheatstone demonstrated that two flat images, like the two in Figure 1.5, fuse in your brain and magically appear three-dimensional. Here was a beautiful example of how the visual system combines 2D images cast on our retinas

A   B

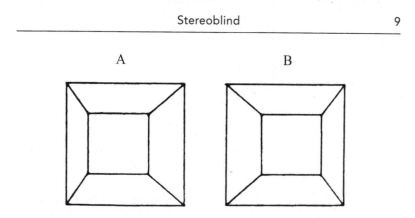

FIGURE 1.5: A stereo pair used by Wheatstone in his stereoscope. (Wheatstone C. 1838. Contributions to the Physiology of Vision.— Part the First. On some remarkable, and hitherto unobserved, Phenomena of Binocular Vision. *Philosophical Transactions of the Royal Society of London* 128: 371–94)

and transforms them into one figure seen in 3D. Shortly after Wheatstone invented his stereoscope, the first 3D cameras were developed. They took photographs from two different perspectives, mimicking the perspectives seen by the two eyes. When these photographs were put into a stereoscope, the scenes appeared in realistic and vivid depth.

Soon, stereoscopes were all the rage in Europe; 3D movies followed in the 1890s and draw great crowds to this day. Just as stereoscopes provide two different views of the world, a 3D movie is made by filming the scenes using two cameras taking pictures from slightly different perspectives. When you put on your 3D glasses in the movie theater, each eye sees the pictures shot by only one of the cameras. Your brain does the rest of the work, fusing the two images into one scene seen in depth. The View-Master, a common toy sold even in supermarkets, works in the same way: it presents to each eye a scene drawn or photographed from a slightly different point of view.

The stereoscope and the View-Master make it easy to fuse two images by allowing each eye to see only one member of the stereo pair. But many people are able to bring images together without the help of a stereoscope by either crossing their eyes or looking "through" the page. See if you can "free-fuse" the right and left pictures in figure 1.5 and get the inner square to recede into or pop out from the page. If you fuse the two images by crossing your eyes, the center square will pop out, while if you fuse the images by looking "through" the paper, the center square will recede behind the outer one.

As a child, I had always wondered why other people seemed so entertained when they looked through a View-Master. I didn't see Disney characters or Superman popping out at me through the toy. All I saw was a flat photograph. Now, in college, I understood at least theoretically what other people experienced. But could I actually imagine what they saw? My newfound knowledge made me wonder if people could imagine a quality, a sensation, that they have never experienced. I thought about people who were totally colorblind. They see no colors at all but live instead in a black, gray, and white world. Could they imagine what the color red looks like? What if they knew all about the science behind color vision? With this knowledge, could they see in their mind's eye what they couldn't see in the real world? I wanted to know the answer to these questions, but I didn't think that I could ever find out. From all that I had read and learned in class about stereovision development, it was not possible for me, cross-eyed since early infancy, to gain stereopsis as an adult.

Since the mid-1900s, the scientific and medical communities have cited strabismus and a related disorder called amblyopia (commonly referred to as lazy eye) as classic examples of developmental disorders that cause permanent changes in vision if they are not corrected within a critical period in early life. These conclusions were based in part on experiments by David Hubel and Torsten Wiesel at Harvard Medical School, the same experiments on cats that my professor had told us about in lecture.

Like the cats in the vision experiments, I'd had misaligned eyes since infancy. If my eyes had been straight and looked at the same object, then neurons carrying information from each eye would have delivered the same input to binocular neurons in my visual cortex. Since my eyes were not straight and saw different things, the binocular neurons in my brain received conflicting input. This situation set up a competition between my two eyes, and for each neuron, one or the other eye won out. Each neuron in my brain now responded to input from only one eye. This change most likely happened during my first year of life, and my eyes weren't cosmetically straightened until I was seven. By this age, the critical period had closed, and my brain was wired in a way that prevented stereovision. While reading in college about critical periods in vision development, I had to conclude that it was too late for my vision to change.

Yet, much more recent scientific research indicates that the adult brain may be more "plastic," or capable of rewiring, than previously realized. The circuitry in parts of our brain changes throughout life as a result of our actions and experiences. With the relatively new science of brain imaging, scientists can now observe changes occurring in people's brains as they learn something new. If you learn to read braille even as an adult, the number of neurons in your brain that receive touch input from your reading index finger increases. Violinists use their right hand for bowing and their left hand for fingering the strings. When playing the violin, the fingers of the left hand move more independently than those of the right. In the 1990s, scientists studied the brains of violinists with magnetic source imaging, and they found that more neurons in the motor cortex of violinists were devoted to the control of the fingers of the left than the right hand.

Indeed, twenty years after college, I witnessed an amazing example of brain plasticity when my husband, Dan, returned from his first space shuttle mission. When Dan and I first started dating in 1976, he told me that he had applied to be an astronaut. I didn't think his goal was realistic, but Dan persisted, and in 1992 he was admitted to NASA's astronaut corps. Four years later, he flew on his first space shuttle mission.

When astronauts orbit the earth on the space shuttle, they are in "free fall": their spaceship and the objects around them are all falling toward the earth together. They don't crash into the earth because the space shuttle is orbiting the earth at just the right speed to continually miss the planet and circle it instead. The astronauts and the objects aboard the shuttle all appear to float. Dan tells me that flying is fantastic, more exciting than hang gliding or bungee jumping, but life in free fall does have its problems. If you close your eyes, you have no sense of up or down. For us earthlings, "down" is toward the center of the earth. For orbiting astronauts, "down" is purely subjective. It could be the floor of the space shuttle or it could be where your feet are placed. As an astronaut, you have to construct your own sense of up and down.

While Dan was on his first mission, our daughter, Jenny, who was ten years old at the time, decided that when he returned she would complete her science fair project on his recovery from spaceflight. On his first full day back from space, Jenny asked Dan to close his eyes and extend his arm straight upward. He extended his arm only about 60° up from the horizontal, while the rest of us, even with closed eyes, could easily judge the vertical and raise our arms straight up in the air. When Jenny asked Dan to stand on one foot with his eyes closed, he immediately fell over. And when she asked him to close his eyes and walk in a straight line along the path from the bedroom door to the bathroom door, he veered off at a 30° angle and crashed into a bookcase. We all found this hysterical—to see our space hero so discombobulated.

You might think that Jenny's experiments indicated that Dan's ability to sense and move had degenerated while he was in space. Instead, Dan had adapted to a radically new environment, the microgravity of outer space. When he first returned to earth, he still acted as if he were moving in space. In the free fall of outer space, the sense organs in the inner ear, the vestibular organs, no longer function normally. If you tilt your head down while here on earth, you know that you have tilted your head because your visual world appears to move up as your head goes down. Your sensory receptors in the neck muscles report that your neck has flexed, and your vestibular sensors in your inner ear tell you that your head has moved toward the earth. When Dan was in outer space, his visual system and the sensors in his neck muscles reported the "downward" movement of his head, but his vestibular sensors did not. He had to find a way to adapt to this sensory conflict if he was going to know where he was and how he was moving in space.

How did Dan cope? Just as the brain merges input from two eyes into one picture, it combines all the input coming from a host of sensory organs into one unified, coherent view of the world. You don't see your friends' lips moving and then hear their voice. You don't see the color of their lips and then their shape. All of this information comes to you in one instant. Yet, when Dan was in space, the information from his vestibular system didn't fit with the information from his other senses. Within a few hours to a few days of moving around in microgravity, he had learned to ignore the conflicting vestibular input and pay more attention to visual information. He adapted: he could fly through his new environment with joy and ease. It's a rush. And what a rush! Despite what you have read in science fiction novels or seen in movies, there are no antigravity rooms on earth. On the space shuttle, Dan learned to cope with an environment that can never be experienced on earth. Dan, in a matter of days, had reorganized the way he sensed and interpreted the world.

Within three days of his return to earth, Jenny discovered that Dan was back to normal. He had regained his balance and his sense of up and down. What's more, after his third spaceflight, a twelve-day mission in 2001, Dan no longer had trouble readapting to earth's gravity at all. After just three trips into space, he had developed two ways of being. He had a system for sensing and moving in outer space and a system for sensing and moving on earth, and he could, within hours, switch from one to the other. Of course, we didn't evolve to live in the free fall conditions of orbiting spacecrafts, but we do have the capacity to adapt to a changing environment, even an unearthly one.

When, in the 1990s, I witnessed how Dan changed the way he sensed and moved after spaceflight, I wondered again if I could change the way I saw the world. In order for Dan to adapt to spaceflight, he had to interact with his new environment. While floating about and handling objects, he learned how to move efficiently in the microgravity of outer space and, in so doing, modified his own brain circuitry. Similarly, studies of the brains of braille readers and violinists indicate that their own actions and habits influence the neuronal connections in their brains.

Yet, according to conventional wisdom, my visual deficits could be explained without involving any physical action on my part. Since my misaligned eyes saw different things, they competed for input onto visual cortical neurons, and on each neuron, one or the other eye won out. My vision, my particular brain wiring, could be explained by considering only the interactions between my eyes and visual cortical neurons. According to this way of thinking, there was no need to consider the way I used my eyes in everyday life and how this might influence my visual circuitry.

But if our actions and habits reshape our neural circuits, then perhaps my own visual habits had influenced my visual wiring. I did not move or use my eyes the way most people do. Since my way of seeing allowed me to move with reasonable confidence and accuracy, my visual habits became entrenched. Did my own actions and habits, and not just local sets of neurons competing for synaptic connections, play a role in shaping my visual brain? Perhaps I could modify the circuitry in my visual cortex by creating experiences that required me to change my way of seeing. Perhaps I could learn to see in 3D.

With these thoughts in mind, I went back to the scientific literature on visual development that I had first studied in college. I wanted to see if the new excitement about brain plasticity had been applied to the treatment of strabismus. But even at the turn of the twenty-first century, the latest papers and books, though full of evidence of the adaptability of the adult brain, still didn't question the critical period in relation to stereovision.

Had I looked at papers and books written by a small subset of optometrists, I would have encountered clinicians who had developed procedures to rehabilitate people's vision, even the vision of individuals like me with a lifelong strabismus. Unfortunately, like most scientists, I had never heard of these optometrists or their work. So, after revisiting the research on critical periods and strabismus, I came to the same old conclusion. If you had asked me in 2001 if I could gain stereopsis, I would have told you that there are limits to how much an older brain can change. A person whose eyes were crossed since infancy would always be stereoblind. And I had been cross-eyed since before I could remember.

2

MIXED-UP BEGINNINGS

"What is REAL?" asked the Rabbit one day, when they were lying side by side near the nursery fender, before Nana came to tidy the room. . . .

"Real isn't how you are made," said the Skin Horse. "It's a thing that happens to you. When a child loves you for a long, long time, not just to play with, but REALLY loves you, then you become Real." . . .

"Does it happen all at once, like being wound up," he asked, "or bit by bit?"

"It doesn't happen all at once," said the Skin Horse. "You become. It takes a long time."

—*The Velveteen Rabbit, or How Toys Become Real,* by Margery Williams

"**S**top wandering, stop wandering."
My parents whispered this to me over and over again when I was growing up. It was our coded message to tell me that my eyes were crossing, and it was time for me to stop daydreaming, pay attention, and straighten my eyes.

My parents first noticed my misaligned eyes when I was only three months old. They consulted our pediatrician, who told them that it was difficult to diagnose strabismus at such a young age due to an infant's wide, flat nose and the folds in the corners of a baby's eyes. The year was 1954, and many pediatricians were unaware then that crossed eyes in infancy could lead to a lifelong loss of stereovision. Since some children with crossed eyes straighten them spontaneously, the doctor suggested that my parents wait to see if I outgrew the condition.

By the time I was two, my eyes were still crossed. We had just moved to Connecticut, where my aunt suggested that my parents bring me to see Rocko Fasanella, a highly regarded ophthalmic surgeon and the chief of ophthalmology in the Department of Surgery at Yale New Haven Hospital. Dr. Fasanella set up his first office in a room on the ground floor of my aunt's father's New Haven house. This provided us with an easy introduction to this well-respected surgeon, and my parents felt very fortunate to be able to put me under his care.

During our first visit, Dr. Fasanella confirmed my parents' suspicions about my eyes. He diagnosed my condition as "constant, alternating esotropia." Alternating esotropia? My parents thought the two words combined had a bouncy rhythm to them, but they were still an ugly-sounding pair. Actually, "eso" is derived from the Greek and means "within," while "tropia," also from the Greek, means "turning." When I looked, or "fixated," with one eye, the other eye drifted toward my nose. I switched fixation from one eye to the other, which made me an "alternating esotrope." And since my eye turns were always present, my esotropia was also constant. Had I turned my "nonfixating" eye outward instead of inward, I would have appeared "walleyed" instead of cross-eyed. I would have been diagnosed with exotropia instead of esotropia. As Dr. Fasanella explained to my concerned parents, 4 to 5 percent of children develop strabismus in either form, but the vast majority of affected infants are esotropic like I was.

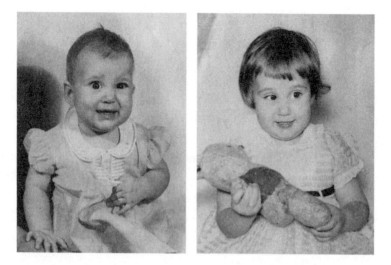

FIGURE 2.1: Early photos of me as an infant and toddler. In each photo, I am looking out of one eye while the other eye is turned in. (Barry family photos)

During each visit, Dr. Fasanella had me look at a penlight with my left eye while he covered my right eye with an occluder, a tool that looked like a large, flattened spoon. While I was looking at the penlight, he moved the occluder from my right to left eye and observed that my right eye straightened to look at the penlight while my left eye, behind the occluder, turned in. If he covered the right eye again, then the left eye moved out to look at the penlight while the right eye behind the occluder turned in. I could straighten whichever eye was doing the looking. Like most cross-eyed children, I had a nonparalytic form of strabismus: I had no problem moving each eye in its orbit. My muscles were functioning fine, but the coordination of the two eyes was off.

My parents were relieved to have a definite diagnosis and asked Dr. Fasanella if problems with my vision explained why I was such a difficult toddler, prone to crying and pounding my head against the floor. He told them that my depth perception was poor—but added nothing further. Doctors knew much less about vision in the 1950s than they do now, and the idea that crossed eyes and other binocular disorders might affect child development wasn't given serious consideration. Had I been born today, my parents may have received more detailed answers.

————

Babies are born with their eyes wide open, but what do they see? The great psychologist William James wrote that infants are born into "blooming, buzzing confusion." However, research on infants indicates that the world may not be as confusing to infants as James imagined. A baby's sensory world is actually very different from an adult's. At birth, the fovea of the retina is not fully developed. Since this area is responsible for our sharpest vision, infants don't see with the kind of clarity experienced by an adult. Newborn babies do not have the ability to focus their eyes at different viewing distances. Instead, they see well only at a distance of about nine inches, the right distance to see their mother's face while nursing. But newborns have some innate perceptual skills—babies, at just nine minutes old, exhibit a preference for looking at a human face.

In addition, the eyes of a very young infant are not always straight. They may wander out of alignment several times a day, but these misalignments do not usually lead to strabismus. A habitual crossing of the eyes emerges at its earliest at about two to three months of age. The early misalignments may result from the baby's attempts to move her eyes into position for viewing at different distances, not just at the distance of her mother's face while nursing. Infants start to look at near objects by converging and then at far objects by diverging their eyes. Most demonstrate an ability to make these vergence movements by about twelve weeks.

During this same period, the fovea of the eye is maturing to allow for sharper vision, and the child develops the ability to focus the lens of the eye in order to see clearly both far and near.

Doctors can watch how babies move their eyes, but it is much more difficult to determine what an infant actually sees. Does a baby see in 3D in the first few weeks of life? Since the eyes are separated in space, the view from each eye is slightly different, and the brain takes advantage of this difference to allow us to see with stereopsis. So, to develop stereopsis, a baby's brain must compare the images seen by the right and left eye. The brain needs to know which eye is seeing what.

One way to determine what babies see is to study what babies like to look at. With these "preferential-looking" experiments, scientists have learned when an infant can see in 3D. To establish when the brain starts comparing the images from the two eyes, scientists at the Massachusetts Institute of Technology (MIT) placed Polaroid goggles on healthy babies whose parents had agreed to have them participate in experiments. These goggles were very much like the Polaroid glasses used in 3D movie theaters. The babies were then shown two different screens as depicted in Figure 2.2 on page 22. While wearing the goggles and looking at screen A, the infants saw vertical lines with one eye and horizontal lines with the other. When looking at screen B, the babies saw identical-looking lines with each eye. Then, the scientists determined which screen the babies looked at longest.

Try combining the two images displayed on screen A in the figure by crossing your eyes or by looking through the page into the distance. You'll discover that it's not possible to fuse the two images into a grid of interlocking horizontal and vertical lines. You may see the horizontal lines for a moment, then the vertical lines. Or you may see horizontal lines in some parts of the figure and vertical lines in the other. But you can't see a place where the

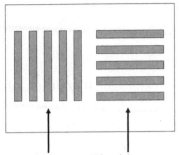

What left eye sees What right eye sees   What left eye sees What right eye sees

Screen A   Screen B

FIGURE 2.2: Preferential-looking experiment. (Adapted from Shimojo S, Bauer J, O'Connell KM, Held R,1986. Prestereoptic binocular vision in infants. *Vision Research* 26: 501–10)

horizontal and vertical lines actually intersect. The images of the horizontal and vertical lines are too different for your brain to combine them together. Instead, you experience "binocular rivalry." Unconsciously, you keep switching attention from one to the other image. While binocular rivalry has been used as a common tool to test perception in laboratory experiments, it's not normally experienced in everyday life by a person with normal vision. We move our eyes so that similar, fusible images fall on both retinas.

Until the babies in the experiment were about four months old, they spent more time looking at the screen that presented horizontal lines to one eye and vertical lines to the other. Then, they suddenly switched their preference to the screen that displayed identical lines to each eye. Prior to four months, the infant brain may not know which input comes from which eye and may combine the right- and left-eye inputs together. After four months, however, the brain determines whether the input comes from the right or left eye, and the babies begin to experience binocular rivalry. Now, the infants look away from the screen with the conflicting vertical and horizontal lines.

Not surprisingly, additional studies have demonstrated that babies do not see with stereopsis until about four months of age. Scientists placed Polaroid goggles on healthy infants and showed them two different stereograms. As with the stereograms that Wheatstone first created, these stereograms presented separate images to each eye. If the two images were identical, then the baby saw a flat picture. If the parts of the image were shifted for one eye relative to the other, then babies with stereopsis fused the two images into one picture seen in 3D, while those without stereopsis saw a flat picture. The babies showed no preference for either stereogram until about four months of age. Then, they spent much more time looking at the stereogram that can be seen in 3D. Scientists have surmised, therefore, that the ability to converge the eyes, to fuse two images together, and to appreciate stereoscopic depth may all develop at about the same time.

These experiments provide some important hints as to when and why crossed eyes develop in infants. Crossed eyes, or esotropia, is likely to emerge during two periods in life. "Infantile esotropia" appears at about two to three months of age, while a second type of strabismus, "accommodative esotropia," usually develops later, at around two to three years. I had infantile esotropia; my eyes began to cross when I was three months old. My parents didn't know what to think at first because sometimes my eyes looked straight and at other times they didn't. Then, over the next couple of months, my eye misalignment increased and became constant. My eye crossing became habitual because I wasn't able to fuse images from the two eyes.

There may be multiple causes of a poor ability to fuse and the development of crossed eyes. Strabismus may be the result of birth trauma, a high fever in early infancy, a slight misalignment in the position of the two eyes in their orbits, a slight imbalance in the strength of the various eye muscles, or abnormalities in visual development. In some cases, strabismus runs in the family. But, as my parents noted, a slight misalignment between the eyes may lead to an even more significant and more constant eye turn over time. My poor ability to fuse images from the two eyes made it hard for me to know where objects were located in space. Crossing my eyes was actually a way to get around this problem.

Infants learn about space through vision, touch, and their own movements. In a classic study, scientists at MIT showed that accurate reaching in cats develops only when the kittens are able to watch their own limbs as they move them. Similarly, human babies develop binocular vision and stereopsis at the same time that they begin to swipe at objects with one hand. By three to four months of age, they often bring their two hands together at their midline. This activity may seem simple, but it is actually a very effective way to start learning about space. As babies bring their hands to their midline, they see their hands move in front of them and watch and feel each hand touching the other while it is also being touched. At four or five months of age, babies will move their hands directly toward a toy to grasp it. Later, as they crawl and then walk, they expand their sense of distance, volume, and space. Their developing visual skills reinforce their emerging motor skills and vice versa.

If I couldn't fuse the images from my two eyes as an infant, then I had double vision (Figure 2.3). Let's say I tried to reach for a tempting toy, but I saw two images of it. Which image of the toy was the real one? Which image could I touch and grab? Indeed, possibly more troubling even than double vision is the phenomenon of visual confusion, an experience of seeing two spatially separated objects in the same location in space.

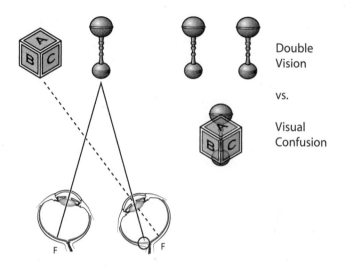

Double Vision

vs.

Visual Confusion

FIGURE 2.3: The left eye is aimed at, or fixating, the rattle while the right eye is turned inward and fixating the toy block. This situation can lead to double vision (the rattle is seen twice) and to visual confusion (the rattle and block appear to be located at the same point in space). F refers to fovea. (© Margaret C. Nelson)

To get a sense of what visual confusion is like, take a small mirror and hold it up, let's say, to the right side of your nose. The reflective surface should be facing away from your nose. With your right eye closed, look straight ahead at an object with your left eye. Then, close your left eye, open your right, and angle the mirror so that a view of another part of the room is reflected into your right eye. Now open both eyes. You should experience the confusing sensation of the two views superimposed upon one another.

As a baby, my eyes were not pointing to the same place in space, so I suffered from visual confusion. I saw two small toys, one viewed by the right and one by the left fovea, as located in the same place even though they were separated in space. Since I experienced both double vision and visual confusion, I had to find a way to adapt to a very mixed-up view of the world.

I learned to adapt to my confused circumstances by suppressing the input from one eye. In this way, I could see a single view of the world. I developed this way of seeing so early that I was rarely aware of seeing double. However, people who develop strabismus later in life may be continually plagued with double vision, and their reports of what it is like have given me insights into what I experienced as a baby.

One of my students, Sarah Merhar, developed a form of strabismus when she was five years old and began to experience constant double vision in high school. I wanted to know how she coped, so one day over coffee, I asked Sarah what her view of the world was like. She said, "I see two images but only one is real. I can be driving and see two images of a car, but I know which one to steer around."

This sounded bizarre, not to mention dangerous, so I asked her how she knew. Sarah thought about this for a moment as she tried to put into words what came automatically to her. Then, she said that the car image seen by her right eye was in context. In other words, the right-eye image of the car was located relative to other things in her surroundings. To Sarah, it had a defined location in space. She added that in class she might see two images of me at the blackboard, but, again, only the right-eye image was real. I was curious: "Does my voice come from the image of me seen by the right or left eye?" Sarah hesitated but said that the voice came from the image seen by the right eye.

"If I were to shake hands with you," she added, "I could reach accurately for and shake only the hand seen by the right eye. If I tried to use the left-eye image, I would miss your hand. It's not that I think about this. I just reach for your hand automatically using the right-eye image."

I pointed to the coffee mug in front of her. "From which of your two coffee cup images does the delicious coffee aroma come from?"

"Now that you mention it," she said, "it only comes from the cup image seen by the right eye."

Indeed, an important function of the brain is to integrate the information from all sensory input into a perceptual whole. For a person with normal vision, images from the two eyes are combined seamlessly into one and associated with other physical characteristics of the object. For Sarah, only the image from the right eye was associated with a defined location in space. This was the image she could accurately touch or manipulate. If she tried to climb the stairs while using only her left eye, she would tumble. If she tried to hammer a nail as seen by only the left eye, she might hit her hand instead. As she learned over time to rely almost entirely on the right eye, she imbued the right-eye image with nonvisual, physical properties, like sounds and smells. Only the right-eye image was "real" to her, and this influenced her entire behavior.

Sarah's descriptions may sound uncanny, but you can replicate her experience by looking at a distant target with your arms at your side. While gazing far away, bring the index finger of one hand a few inches from your face. Keep looking in the distance but be aware of your finger. The fact that you see two images of your finger is perfectly normal. Scientists call this phenomenon "physiological diplopia." When you look into the distance, your eyes diverge so that the distant target casts an image on the fovea of both eyes. As a result, a near object, such as your finger, casts its image on distant noncorresponding points on your two retinas and is seen as double. Rarely are we aware of these double images, but under some conditions we can bring them to our attention.

Now that you see two images, you can experience the real/
unreal dichotomy that Sarah faced when she saw double. While
maintaining your gaze in the distance, take your free hand and try
to touch your finger. Which image can you touch? For most of us,
only one of the images is solid and graspable. If, like Sarah, you
always saw double, then over time, you would associate sounds
and smells from a given object with only one of the two images.
When Sarah reached for an object, she reached for the "real" image. This is what happened to me when I was an infant. But I developed strabismus at such a young age that the second image
actually faded from consciousness; I no longer saw it at all.

To make it easier to disregard one eye's input, I turned in the
eye that was not doing the looking. With my eyes in this position,
an object would cast an image on the fovea of my fixating eye and
on a nonfoveal region of the turned eye. As a result, the image seen
by the fixating eye would appear clear, while the image seen by the
turned eye would have less definition and detail. Under these conditions, it was easier to discount the displaced image from my
turned eye, to regard its image as unreal. The more I turned in my
eye, the less clear the image from that eye would appear, making
it that much easier to ignore.

But why, my parents wondered, did I, like most infants with
strabismus, turn the eye in and not out? Dr. Fasanella could not
give them an answer then, but recent research has shown that
young infants, even with normal vision, can move each eye more
effectively toward the nose than away from it. Thus, if I needed
to move one eye out of alignment in order to suppress its input,
it was easier to do so by turning the nonfixating eye in rather than
out. By crossing one eye, I could discount one image and see a single view of the world. This solved one problem but created another: I had to develop a sense of depth without stereopsis.

Many scientists and physicians have assumed that a cross-eyed
infant can still develop a good sense of depth using a cue called

motion parallax, a way of seeing depth involving movement of the head. But this is not the case. People who have been cross-eyed since early childhood see much less depth using motion parallax than people with normal vision, and this, along with the lack of stereopsis, greatly compromises depth perception.

An easy way to experience motion parallax is to look out the window and slowly sway side to side. While swaying, keep your gaze fixed straight ahead. As you move right, near objects appear to move left while distant objects move with you to the right. The opposite happens when you move left. What's more, near objects appear to move a greater distance than distant ones. The next time you are a passenger in a car, pay attention to the scenery as it rolls by. You'll see that nearby objects appear to move away from you in a direction against the car's motion, while distant objects appear to move in the direction of the car. The way these objects move relative to each other contributes to your sense of depth.

Since stereopsis and motion parallax play a major role in our perception of depth, infants with strabismus or other binocular vision impairments develop a sense of distance and space with an impoverished set of cues. They depend more on "monocular cues" to depth, such as shading and perspective. As a result, many cross-eyed babies show delays in mastering tasks like grasping a toy or holding a bottle, and older children with the same problems may even show abnormalities in gait and posture. Finally, a loss of stereovision early in life leads to a greatly impoverished sense of distance and space. Of course, my parents didn't know any of this when they first took me to see Dr. Fasanella. They knew only that they had a very temperamental two-year-old whose eyes wouldn't stay straight.

After my first visit, Dr. Fasanella prescribed for me my first pair of glasses. They had heavy frames and were actually made of glass, not of the lighter, safer materials used today. My first childhood memory is of sitting on the stoop outside our kitchen feeling the

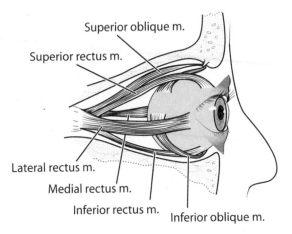

FIGURE 2.4: The six muscles that move the eyes. (© Margaret C. Nelson)

weight of these glasses on my nose and ears. There was a rhodo-
dendron bush to my left that I wanted to look at, but I was afraid
to turn my head for fear that the glasses would fall off and break.
My glasses were bifocals, which made it easier for me to focus on
objects nearby. If I'd had a particular type of strabismus called accommodative esotropia, wearing bifocals might have straightened
my eyes. But even after several months of constant bifocal use, my
eyes remained crossed. So, Dr. Fasanella decided to operate.

My first surgery occurred when I was twenty-eight months old.
Dr. Fasanella explained to my parents that the eye muscles hold and
move the eyeballs in their sockets (Figure 2.4). Think of my eye,
he told my parents, as the head of a horse and the eye muscles as
the horse's reins. Imagine the horse's head as pointing, let's say, to
the right. Shorten the reins on the left and lengthen the reins on
the right and you can straighten the horse's head.

Dr. Fasanella realigned my eyes in their sockets by shortening
the length of some muscles and changing the point at which they
inserted into the eyeball. In my first surgery, he repositioned the

right medial rectus muscle on the eyeball so that this muscle, which pulls the eyes in, was at a mechanical disadvantage. He did this in an effort to decrease my tendency to turn my eye inward. He also repositioned the lateral rectus muscle so that it was more effective in pulling the eye outward. After the procedure, he noted in my records that further surgery would be required on my other eye "because of the large amount of esotropia." Before my first surgery, he had warned my parents that more than one operation would be necessary. My parents trusted him and accepted his conclusions.

A year later, I had a second operation in which the corresponding muscles of the left eye were cut and repositioned in a similar manner to those of the right. As often happens with strabismus, vertical eye misalignments developed over time, and after the first two surgeries, my left eye gradually moved into a position higher than my right. So, when I was seven, Dr. Fasanella performed a third surgery to move the right eye upward. He also cut part of the tendon of the left medial rectus muscle to further weaken its ability to move my eye inward.

I remember my third surgery quite well. My hospital room was long and narrow with two beds, one for me and one for my mother. During my stay there, my mother's friend Eppie came to visit. A nurse at the hospital, she brought me a set of finger puppets that I played with and treasured for years. Something about her was enormously comforting, wise, and warm. Many years later, I learned from my mother that Eppie, or more formally Florence Wald, was dean of nursing at Yale New Haven Hospital and had started the first hospice unit in the United States. My mother remarked that even a seven-year-old child can recognize an exceptional individual.

When I was wheeled into the operating room, a man draped in a long gown came up to me carrying a narrow tube with a horrible stench.

"Do you like this smell?" he asked through his surgical mask.

When I said no, he told me that if I took ten deep breaths, the smell would go away. I eagerly breathed with him but only got to three before drifting out of consciousness. When I awoke in the recovery room some time later, I realized that the tube must have contained the gas that had put me to sleep. What was this business about taking deep breaths to make a smell go away? Why hadn't he simply told me that the gas would put me to sleep? Did he think I wanted to be awake while Dr. Fasanella cut into my eyes? I felt betrayed.

I also awoke with an uncomfortable patch over my right eye, a patch that was changed every day for two weeks while I was confined to my bed at home. But there were compensations. I had always wanted a dog, and after my two-week convalescence, my parents surprised me by taking me to a neighbor's house where they introduced me to a little puppy that became our first family pet.

My parents had such confidence in Dr. Fasanella, and I found him to be such a kind man, that the surgery itself hadn't frightened me. But I worried about going blind. On most nights throughout childhood, I woke up in the dark and checked for the familiar beam of light from a hallway lamp that would shine underneath my bedroom door. I would make sure I could see the light first with one eye, then the other. Once convinced that both eyes were still working, I quickly fell back to sleep.

After the operations, I certainly looked better. My parents and I were very pleased with the results. My eyes looked straight most of the time. Since I could keep my eyes aligned best when I looked upward, I tended to open my eyes wide with my eyebrows raised. I had large eyes in a small head, and this combination, along with the way I positioned my eyes, gave me the look of a startled bug. With my saucer-eyed look, school photos were always a problem. I tried so hard to make my eyes look straight in front of the camera that they ended up looking like they were popping out of my head. When I brought my class photos home, my parents didn't comment on my bulging-eye look. Instead they purchased a set of my pictures, along with the much cuter ones of my brother and sister, and then quietly put all the photos away in a drawer. Kids at school called me "frog eyes," but my parents and their friends told me constantly that my eyes were beautiful. So, despite my unflattering nickname and comical photos, I was happy with my straight, if bulging, eyes.

Even though my eyes appeared straight, I still didn't use them normally. Dr. Fasanella told me that I continually switched from one eye to the other. He called me an "alternator." Had I used only my left or right eye most of the time, I could have lost vision in the unused eye. Since I alternated, I retained good acuity in both eyes. Dr. Fasanella seemed pleased with the overall result, so I felt proud of how well I had come through the operations. Although I was told that my depth perception was a little weak, no one explained to me that I lacked stereovision. My parents weren't trying to hide anything; they simply didn't understand what I was missing. So, I remained ignorant of this fact until that fateful lecture in college.

For me, cosmetic alignment of my eyes did not change the way I used them. I continued to see as I had before the surgery. This is true for many children with strabismus, particularly if they have surgery after the first year of life, the presumed critical period for the development of stereovision. Even though my eyes looked straight, they were not as straight as nature intended them to be. When my surgeon repositioned my eyes in their orbits, he had to be careful not to overshoot and turn me from a cross-eyed child into a walleyed youngster. So, after my surgeries, my eyes were still slightly crossed, although to the casual viewer they appeared normal. Given my former viewing habits, I was less likely to try to combine images from the two eyes and develop stereovision. I simply went back to my old way of seeing.

Like me, many strabismic infants and children require more than one surgery for adequate alignment of the eyes. If a baby's eyes are brought into closer alignment by surgery, yet the baby still can't merge images, then he continues to receive conflicting input from the two eyes. To have clear, single vision, the baby must still suppress one eye's input by turning in one eye once again or moving one eye out of vertical alignment, thereby defeating the results of surgery. Babies who can fuse images and develop stereopsis after surgery are more likely to keep their eyes aligned and require no further operations.

Had I seen a developmental or behavioral optometrist as a child, I would have been given optometric vision therapy. Ironically, at the time of my surgeries, medical doctors, developmental psychologists, and optometrists were working together at the Gesell Institute of Human Development right near the hospital where I had my surgeries. They studied and treated cross-eyed children. Then, as now, eye surgeons and optometrists didn't generally communicate or work with one another, so no one mentioned the Gesell Institute to my parents. If they had, I might have learned how to coordinate my eyes for stereovision and avoided a third operation. Almost certainly, I would have had an easier time in school.

3

SCHOOL CROSSINGS

The most instructive experiences are those of everyday life.
—Friedrich Wilhelm Nietzsche

I dreaded going to grade school. Throughout childhood, I had 20/20 acuity in both eyes, but I had trouble learning to read. When I looked down at the letters on the page, they didn't stay in one place. This problem grew worse as the print got smaller. My reading difficulties came to a head when I performed miserably on a standardized achievement test. These "objective," scientifically designed tests were thought to reveal a person's native intelligence. The tests were far more accurate, many school administrators felt, than the observations of a skilled, experienced teacher, even one who had observed a child for a full school year.

My school divided the children in each grade into four groups, and I began third grade in a class with all of my friends. Although we were not told why we were each assigned to a particular classroom, the groupings were obvious to me and all of my schoolmates. One class was for the above-average students, one for the average learners, one for the below-average pupils, and one for the children with "special problems." On the first day of third grade, I was placed in the above-average class but survived there for only one week. A mistake had been made. My score on the standardized test from the previous year indicated that I was supposed to be in Mrs. Danner's special-problems class.

The assistant principal came into the classroom and asked me to stand up. She instructed me to leave the classroom and asked a boy to drag my desk behind me as I walked down to Mrs. Danner's room. The desk made an awful noise as it scraped along the floor. I felt humiliated and shamed by being made the center of so much attention.

When I got home from school, I was very upset, and my parents arranged a meeting the next day with Mrs. Bell, my teacher from the previous year. I had spent the preceding year in great fear of Mrs. Bell, for she had a habit of tipping over a student's desk if the contents were not kept neat and organized. My parents learned, however, that Mrs. Bell was my great advocate. She had argued fiercely with the principal over my classroom placement. She suggested to him that having my eyes rearranged in their sockets three times in five years had interfered with my reading skills. After talking with Mrs. Bell, my parents met with the principal. But he insisted that the tests were accurate and objective. My aptitude, he told them, was well below average, and I had been wisely moved into Mrs. Danner's class. My vision was not considered to be a factor.

My mother panicked. The most law-abiding individual on earth, she snuck into the school's office after hours and stole a copy of the achievement test on which I had performed so poorly. At home, she took me down to the basement, told me not to breathe a word to my brother or sister, and gave me the test. In the quiet, relaxed atmosphere of my home, I did much better. Again, my mother met and argued with the principal, but she couldn't admit to him that she had snuck into his office and absconded with the test. I remained in Mrs. Danner's class.

I hated all of this fuss about my abilities. I was embarrassed around the kids in my neighborhood because I assumed that they all thought I was dumb. At school, I said nothing, never raised my hand to answer the teacher's questions, and essentially tried to disappear. But I did gain something positive from this experience. I learned that my mother, normally so gentle and soft-spoken, would not only fight but break the rules for me, and she questioned the school authorities—not my intelligence.

Mrs. Danner was unusually calm, patient, and steady, but the children in her classroom were not. There was something wrong with all of us. Some of my classmates had physical problems, most had trouble paying attention, and several were completely disruptive. I made one good friend in the class, a boy name Scott, who had suffered from polio and walked with a limp. Years later, when we were in high school, Scott got into a motorcycle accident and injured his good leg. The high school principal got on the public address system, told the school about Scott's accident, suggested that we all send "get well" cards, and then gave a speech about reckless behavior. I remember my anger building as I listened to him preach. I believed that Scott's "reckless" behavior resulted from the way he had been treated since the first days of school, from growing up labeled as "different" and being regarded as a "failure." As I listened to the principal, I reminded myself never to be too quick to judge other people.

My mother taught me how to read when the school gave up on me. She read with me and to me constantly. Often, she would leave a new book on my bed that I would discover when I returned home from school. I was very shy and felt happiest when exploring the countryside, categorizing wildflowers, trees, and rocks. My mother would leave me books about nature and animals. When I finally discovered Walter Farley's *Black Stallion* series, I was hooked and began to read for pleasure.

By fifth grade, I had become a competent, if slow, reader. I was finally transitioned out of the special-problems class and into a regular classroom. I was a fanatic student, compulsively checking my answers again and again in the hopes that my hard work and discipline would hide my tested lack of intelligence. There was one reading exercise, however, that I simply could not do. My teacher called it "controlled reading," and it involved reading a story while the words moved by on a screen in the front of the classroom. After the text rolled by, I had to answer questions in a workbook relating to the story. I could not follow the moving words and was sure that I would be sent out of the classroom, desk and all, back to the special-problems class. I quickly discovered, however, that the answers to the questions were printed on the back page of the slim workbook. If I pressed down on the page with the questions, I could see through to the back page. It was the only way I could answer the questions correctly, and it was the only time I ever cheated in school.

Common experience tells us that our vision plays a large part in our ability to read and do well in school. Yet, many school administrators and physicians have long questioned the connection between vision and learning. Most of us consider "perfect" vision to mean 20/20 eyesight as measured by identifying the letters on the Snellen eye chart—commonly recognized as the chart with the big letter E on top. Yet, good eyesight (or acuity) and good vision are not the same thing. We need more than 20/20 eyesight to read a book. When we read, we view letters and words positioned about sixteen inches from our face, not twenty feet away, and we must be able to sustain close viewing for long periods. We look at the letters with two eyes, not just with one as in an eye exam, and we have to move our eyes across the line of words in a coordinated manner. Finally and most importantly, we have to extract meaning from the words. All of these processes are involved in good vision and affect our ability to learn.

Although the exact role of vision in learning is a subject of intense debate, many scientific studies support a connection between vision and reading. For example, one paper published in 2007 examined the visual skills of 461 high school students who read at two or more levels below the established level for their grade. Of these students, 80 percent had eyesight of 20/40 or better when looking at an eye chart placed twenty feet away. However, at least one-fifth of these students had trouble focusing on the text for sustained periods. What's more, the majority of the students fell below normal standards in their ability to converge and diverge their eyes for stereovision. Additional papers have demonstrated a correlation between reading skill level and the ability to see with stereopsis.

Why would there be a correlation between stereopsis and reading? After all, you can read with only one eye; when you read, you are looking at a flat page, not a three-dimensional object, and you do not need to judge depth while reading. But poor or absent stereopsis indicates difficulty merging the information from the two eyes. Instead, that information may be conflicting—as was my situation. I had 20/20 eyesight with both eyes and no problem passing a standard school vision screening. Yet, my vision was abnormal because I did not use my two eyes together. The uncorrelated information from my eyes greatly disrupted my ability to read.

For years, scientific research done on eye movements during reading monitored the movements of only one eye. It was assumed that the two eyes moved in concert, so observing the movement of one eye was sufficient to know what both were doing. But recent experiments examining the movement of both eyes together during reading have yielded some important surprises. For most of us, our eyes do not always point to the exact same place on the page when we're reading. For about 50 percent of the time, the right eye is aimed about one to two letters to the right of the letter seen by the left eye. This doesn't present a problem to the reader because the images from the eyes are merged in the brain. The information is combined in a cooperative fashion.

What happens, however, if the two eyes register conflicting information? Since I was cross-eyed, I cross-fixated. When I was learning to read, my right eye saw letters located to the left of the letters that I saw with my left eye. I didn't merge images from the two eyes but rapidly alternated between my left- and right-eye views. Although I am not dyslexic, I distinctly remember being in first grade and trying to figure out whether the word I was reading was "saw" or "was." Pinpointing the exact location of letters on a page was very difficult.

By the time I reached fifth grade, I had unconsciously found a way to read books comfortably (even if I could not do the controlled-reading task.) However, I did not discover how I used my eyes for reading until I was an adult and underwent an eye exam with a developmental optometrist. When I looked at the words with my right eye, my left eye turned in, or crossed, by 15°. If I read with my left eye, the reverse happened. The fact that I turned in my eye by 15° reveals how I eliminated the interference between the two eyes, for our "blind spot" is located 15° from the center of the retina. In this region, where the optic nerve leaves the retina for rest of the brain (figure 1.2), there are no light-sensing cells. When I read with my right eye, the image of the word fell on the fovea of my right eye and on the blind spot of my turned left eye. The reverse held true when I read with my left eye. Unconsciously, I had found an effective way to eliminate the conflicting image from the nonfixating eye.

Examples abound of children who have visual problems misdiagnosed as learning disorders. Indeed, hearing Michelle Dore's story about her son, Eric, reminded me of how common it is to misdiagnose youngsters with vision problems. Michelle's son was determined in kindergarten to have attention deficit hyperactivity disorder (ADHD). Eric's teachers told Michelle that he was smart and could be a good student if he would only settle down. So, Eric was put on a host of drugs intended to control his unruly behavior. His parents even enrolled him in a pilot study for a new type of medication, but the drug made Eric feel like a zombie. When, in school, Eric was slow at copying the information from the board, his teacher assumed he was not paying attention and sent him to the time-out room as punishment. Of course, this punishment only made it harder for Eric to keep up in class. When Michelle saw the time-out room, actually a windowless closet, she took him out of that school and enrolled him in another in her constant search for an environment in which he could succeed.

Eric was not an avid reader, but he liked the Harry Potter books because the print was large with more white spaces between the words. When Michelle observed this, she insisted that the school provide Eric with tests in which the words were printed in larger letters, and gradually his performance in school improved. About this time, Eric developed a passion for hockey, a passion he shared with his dad. His father coached him; he practiced like mad, becoming talented enough to make one of the better teams in the region. Even so, Eric always seemed to be a split second behind in his moves, and he often endured bullying by his peers.

One evening at dinner, when Eric was in his teens, Michelle noticed that one of Eric's eyes would turn out intermittently. Since this happened when Eric was looking closely at an object with his eyes cast downward, she had never observed the eye turn before. She took him to the pediatrician, who noticed nothing wrong. But Eric's mother still suspected that something was holding him back.

When, some time later, Michelle saw a flyer advertising a new clinic to treat ADHD, she took Eric to it. Among the tests administered at the clinic were several that examined eye coordination. The tests revealed that he had trouble moving his eyes together when reading. Encouraged by this new piece of information, Michelle took Eric to see a developmental optometrist.

After all these years of doctor's visits and school tests, Michelle finally learned that Eric's difficulties resulted from a visual condition called convergence insufficiency, a common but often undiagnosed cause of reading troubles. Eric aligned his eyes appropriately for far viewing but not for near. When an object was brought close to his face, Eric abandoned stereoviewing. Instead, he looked at the object with just one eye and allowed the other to wander out just like he had at dinner that night. If he couldn't converge his eyes for near viewing, Michelle wondered, how had her son managed to read a book in school or judge the distance of other hockey players on the ice? How had he followed an approaching puck or caught a pass? As Michelle thought about her son's visual challenges, she felt a renewed appreciation for his determination and perseverance.

As Michelle discovered, convergence insufficiency is not easy to spot without careful testing. What's more, most children don't realize that they have a vision problem. They don't know that the words shouldn't jump around on the page when they read. They don't realize that the letters shouldn't appear doubled or blurred. Of course, these unrecognized difficulties with reading and the subsequent troubles with schoolwork can lead, as they did in Eric's case, to behavioral problems. Since standard school vision screenings do not pick up this vision problem, children with this condition who appear restless in school and are poor learners may be labeled with ADHD or other disorders.

According to a recent National Eye Institute study of 221 children, the most effective treatment for Eric's condition is a combination of office-based and home-based vision therapy. After five months of vision therapy with a developmental optometrist, Eric's reading and grades improved, and he was able to realize his dream of competing in a highly competitive hockey league. With renewed confidence, Eric entered college and became one of the top students in his class.

Every person with a binocular vision disorder copes slightly differently. I found a way to adapt my vision for reading by fifth grade, but my vision led to other challenges throughout school. When I was in eighth grade, all of the boys had to take a shop course, and all of the girls had to learn to sew. I absolutely loathed sewing. When I was instructed to sew by hand, I found it extremely tedious and tiring because it was difficult to keep my gaze on the needle as I moved it in and out of a piece of cloth. Using the electric sewing machine was even worse. I couldn't follow the rapidly oscillating sewing machine needle in order to guide a piece of fabric underneath it. To my dismay, I was supposed to make a dress and then model it in a fashion show at the end of the school year. My mother didn't sew, so she couldn't help me, and my grandmother, who still used a manual sewing machine with a foot pedal, saw no reason for me to learn to use an electric model. Finally, my father, an artist, seeing my distress and total incompetence, did my sewing projects for me. After getting through sewing class, I never touched a sewing machine again, and when I got married, I told my husband that he would have to hem his own pants.

To read you need to pinpoint the location of letters on a page, to sew you need to position the needle on the cloth, but to get around you need to know where your own body is in space. My favorite way of getting around was on my own two feet—by walking or running. I felt very unsteady on a bicycle, and during the first year after learning to ride, I walked the bike down hills. I did not want to learn to drive. When I first got behind the wheel, the oncoming traffic petrified me, and I thought that good driving basically involved avoiding head-on collisions. Since I concentrated on looking straight ahead and paid little attention to my peripheral vision, cars and pedestrians seemed to appear suddenly out of nowhere. It was hard to maintain a steady gaze while looking down a long straight road. The distant scene seemed to grain out like static building up on a TV screen. I would shake my head to get back a clear view. Not surprisingly, I got my driver's license a year after most of my peers. On my first solo outing, a three-mile trip to my boyfriend's house, I scraped the side of the car on a stone wall on one side of his driveway. My parents said nothing but offered to drive me anywhere I needed to go.

Indeed, I was always just a step behind everybody else. In college, I wanted to join an athletic team but was not good enough to participate in softball or tennis. Since I loved the water, I joined the women's crew team and rowed in an eight-person shell. In a crew boat, the rowers sit in a line facing backward and match the rhythm of their strokes with that of the person sitting in the stern. I was not particularly surprised when my coach told me that my pace was just a little slower than everyone else's. Happily for me, she didn't drop me from the team. Instead, she put me in the bow of the boat so that my slightly delayed stroke wouldn't confuse another rower sitting behind me.

Obviously, I wasn't going to become a star athlete, but I did enjoy my science classes and liked being out of doors. So, in college, I thought I would study animal behavior and follow my keen interest in birds. One of my professors was a great birder. Like many naturalists, he had uncanny powers of observation. "Driving and birding are my two great loves," he would say as he accelerated the car to dangerous speeds on our way to prime birding sites. I thought his two great passions made for a strange combination until I realized that his driving style was matched only by his quickness in observing the natural world. He had sharp eyes and ears and would spot birds in the foliage long before anyone else. Despite my intense interest in ornithology, I soon noticed that I was always the last in the group to see the birds. On one trip to Hawk Mountain in Pennsylvania, I was supposed to help spot and catalog the migrating hawks. I wasn't much help to the other birders, often missing the majestic birds entirely. On that trip, I decided that I would not make a very good ornithologist. I would always enjoy watching birds, but I didn't see quickly enough to make important, new observations.

Even if I was poor at observing animals in nature, I could study the neurons that help them sense and move through the world. So, I took my first class in neurophysiology and in that class made my first recordings of nerve cells. When neurons fire, they generate electrical signals that resemble the waveforms seen in an electrocardiogram. I loved watching these signals; it was like listening to the mysterious language of nerve cells. I wanted to learn more about this language, but to do so, I realized, I would have to overcome a new visual challenge.

To record the electrical activity of individual neurons, I had to use a stereomicroscope to see and then position a tiny glass needle through the thin outer envelope of a nerve cell. The stereomicroscope, with its two eyepieces, was designed to help people see the needle and nerve cells in three dimensions, but by this point I knew that I didn't have stereovision. I decided not to tell my teachers initially about my visual deficit lest they discourage me from pursuing my interests. Instead, I developed my own strategy to reconstruct the three dimensions. I continually adjusted the fine-focus knob on the microscope, which brought my view of the electrode and nerve cell in and out of focus. By looking at the changing focus, I could gauge the depth. My lack of stereovision was less of a problem when using stereomicroscopes than it was when driving cars or watching birds because I didn't have to see things as they sped by. Instead, I could take my time and move deliberately. I used this microscope technique all through college and graduate school and throughout much of my career as a neurobiology researcher and professor.

However, as often happens with people with childhood strabismus, my vision began to trouble me more in my late thirties.
By this time, I was married with children and working as a professor at Mount Holyoke College. When I was forty years old, I
went to see a local, well-respected ophthalmologist. My lack of
stereovision was not what concerned me. I had long ago accepted
that I would never gain stereopsis. Instead, I told the doctor that
the world appeared jittery, especially when I looked in the distance. I had trouble reading road signs when driving because I
couldn't keep my eyes on the words. I would slow the car to a
crawl in order to read them, unnerved by the honks from angry
drivers behind me. If I had to go someplace new, I would head out
the day before at the least-trafficked time and drive to the new location so that I wouldn't get lost on the following day. What's
more, I told the doctor, I could not see clearly from the middle
of my children's school auditorium. The faces of the children on-
stage were all a blur.

The ophthalmologist checked my right eye and then my left
and told me that I was very lucky. Without glasses, my right eye
saw 20/50 while my left eye saw 20/60, and my glasses totally corrected for these small flaws. "Why, then, am I having trouble seeing?" I asked him. He dismissed my concerns. "There's nothing
wrong with your vision," he insisted, adding that if the world appeared unstable to me, it was probably because I'd been trauma-
tized by my childhood surgeries and, as a result, "dreamed up" my
vision problems. Perhaps, he said, I needed to see a psychiatrist.

I left his office as soon as possible after grudgingly paying the
bill. I felt angry and tearful, but I told myself that, in the end, the
news was really good. The doctor didn't say something like, "Your
complaints indicate that you have an incurable brain tumor." He
told me that my eyesight was fine. I had no vision problems. If I
was simply less anxious, these troubles would surely go away.

4

KNOWING WHERE TO LOOK

You never really understand a person until you consider things from his point of view.

—Atticus Finch in *To Kill a Mockingbird*, by Harper Lee

In his thoughtful and moving memoir, *Touching the Rock*, John Hull recounts what it is like to be blind. He describes sitting on a park bench while listening to the sounds around him. He hears children shouting, people running, balls bouncing, wind blowing, birds calling, and traffic rumbling. It is "an astonishingly varied and rich panorama of music, movement and information," he notes, and yet he adds that things that are beyond his reach and make no noise simply do not exist. Only when his children shout to him from a paddleboat floating on a nearby pond does he know that they are there.

John Hull can get up from the park bench and discover more about his surroundings by using his cane and sense of touch. Yet, most of us can find out where we are in the world much more quickly; we simply look around. Vision allows us to be active participants in our world, continually moving through it and molding it to our needs and our desires.

Indeed, our ability to move confidently and accurately is intimately tied to our ability to see. If we didn't have legs and arms

FIGURE 4.1: The tunicate as a sessile adult (left) and swimming larva (right). (© Margaret C. Nelson)

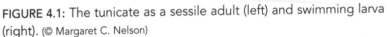

for climbing trees and fingers for manipulating objects, we would never have needed such a complicated visual system or brain. Similarly, our marine cousins, the tunicates (Figure 4.1), have eyes when they are young larvae in order to swim about and find a good place to settle down. Once they come across a suitable rock on the sea floor, they attach themselves to it and transform from active swimmers into stationary filter feeders. At this point, their eyes and brain degenerate as they have little need for a good visual system and brain if they are not moving.

To move accurately, we must first move our eyes. Imagine that you are driving down a treacherous mountain road. You gaze directly at the curve a second or two before rounding the bend, then turn your head in the direction of your gaze and masterfully steer the car around the curve. On a more pedestrian level, when you walk down the street and turn the corner, the rotation of your body follows the direction of your gaze. You reach for an object first by directing your eyes, then your head, and finally your arm toward the target. The movement of your eyes anticipates and then directs the movement of your head, body, and limbs.

We move our eyes before our body because we must look directly at objects to see them clearly and in detail. If you try to read this book by holding it to the right or left of center while looking straight ahead, you will still be able to see that there are letters on the page, but you won't be able to identify them. You do not have to move the book very far from your direct gaze before you can no longer read it.

We must look directly at the words because our sharpest, most acute vision is located in the central part of the retina. This region, called the fovea, receives input from about the central 2° of the visual field, an area about equal to the size of a quarter as seen from arm's length. If you try to read this book while moving it away from your central gaze, you will begin to have trouble recognizing the words when the letters move out of foveal range. Since the fovea sees such a small part of the visual field, we move our eyes constantly to see one part of a scene and then another in sharp detail.

To pick up an object or move around it, however, not only must you see it clearly, but you must also know where it is in space. To do that, you need to know the direction in which your eyes are pointing. If we had only one eye, this would be simple. Assuming that your head was oriented straight forward, you could determine the location of an object by where your eye was aiming. But with two eyes, the situation is much more complex.

While holding this book about fourteen inches from your face, look at the large X below while thinking about the position of your eyes.

X

If the X is directly in front of you, your right and left eye have each turned in to the same degree but in opposite directions, as in the picture of the two eyes shown in Figure 4.2.

Your sensed visual direction—the direction in which you are looking—is not the direction in which either eye is pointing but one that seems to emanate from the center of your forehead. Since our eyes are both aimed at the target, the image of the X is cast on corresponding retinal points. For most people, images that fall on corresponding retinal points appear in the same subjective visual direction. In other words, we interpret these right- and left-eye images as emanating from the same point in space. We see one X in one direction, in this case, straight ahead.

Most of us interpret objects that fall simultaneously on the fovea of both eyes as located in the same place in space. In tech-

FIGURE 4.2. (© Margaret C. Nelson)

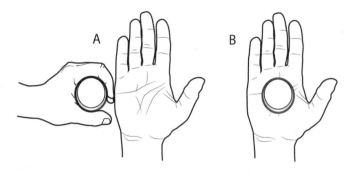

FIGURE 4.3: Parts A and C demonstrate the setup for the experiment, while part B shows what a viewer with normal vision will see. (© Margaret C. Nelson)

nical terms, we see with "normal correspondence," a phenomenon illustrated by the following "hole-in-the-hand" experiment (Figure 4.3). Take a piece of paper and roll it into a tube. Close one eye and then look at a distant object through the tube. Place your free hand next to the tube with your palm facing toward you. Now open the closed eye and keep looking in the distance.

Most people experience the weird impression that there is a hole going right through their hand. This strange illusion occurs because our brain merges the images from the two eyes by combining inputs that fall on corresponding regions of the retinas of both eyes. Since one eye registers the image of a tube on the retina where the other eye registers an image of a hand, we interpret the tube and the hand as coming from the same point in space and see the tube as running through the hand.

One of my friends, nicknamed Red Greene, discovered for himself the intimate relationship between where his eyes were aiming and where he located objects in space. Although Red was a housepainter who climbed ladders and painted houses for many years, he was amblyopic: since early childhood, he had had very poor vision in his right eye. For most of his life, the world seen through his right eye looked like the view through the frosted glass of a shower door.

When Red was in his fifties, the vision in his left or "good" eye deteriorated, a situation that could have proved dire except that acuity improved dramatically in his weak right eye. In fact, his right eye now has 20/20 acuity. While this turn of events may seem surprising, similar cases have been documented in which the amblyopic eye regains vision following severe impairment of the "good" eye.

However, Red had a very hard time transitioning from seeing predominantly with the left eye to seeing with the right. His world was unstable. Fluorescent lights bothered him. Reading was almost impossible. He became nervous and anxious around crowds and in busy environments. He didn't last a single day at Disneyland. Red consulted his local ophthalmologist and traveled to Boston to consult with a neuro-ophthalmologist at the New England Eye Center of Tufts University. Even though Red went to an excellent center for vision research and care, no one could figure out how to help him or even why he was complaining. After all, he had perfect eyesight in one eye. The best they could do was to suggest that Red wear a patch over his left eye and use only his good right eye for seeing. When this didn't work, they suggested that he add a black-tinted lens to his glasses to further reduce the sight of his left eye. He resigned himself to the fact that nothing more could be done.

But Red's wife, a nurse, learned from her colleagues that a local optometrist treated many stroke patients and might be able to help. During Red's first visit, the optometrist evaluated how he used his two eyes together and noticed something that none of the other doctors had. Although his left eye was no longer functioning well, it was still his dominant eye; for almost all of his life, his brain had regarded images from his left eye as the "real" images. His left eye directed where Red would look and move, a real disadvantage to him now that only his right eye saw clearly. What's more, because he was amblyopic, his two eyes did not work well together. The optometrist put prisms in Red's eyeglass lenses that aligned the visual fields of his eyes. Immediately, Red felt a calmness come over him. He now saw with a steady gaze and could read comfortably. He got up from his chair and marched around the examining room in happy disbelief. Red's vision and life were transformed.

Red's difficulties highlight the problems with spatial localization that arise if your eyes are misaligned. If you can't direct both eyes to the same point in space, the two images of a single object do not fall on corresponding points of your retinas. How do you interpret the unique location of an object when your two eyes present conflicting information? Throughout most of his life, Red coped with this issue by discounting, or suppressing, the input from his right eye. He used only his left eye to locate objects around him. Suppression does indeed solve the spatial-location problem, but it creates another.

When we suppress the input from one eye, we lose up to half of the visual information that's available to us. So, some people with strabismus unconsciously develop an additional strategy that allows them to make simultaneous use of both eyes, even though the eyes are looking at different regions of space. To do this, their brain abandons the idea that the two foveas point to the same

location in space. Instead, they aim one eye at a target and interpret the location of images that fall on the central retina of that
eye as straight ahead. (Assume here that the head is turned nei-
ther right nor left but is oriented straight forward.) Though aware
of images that fall on the fovea of the strabismic, or turned, eye,
they localize these images with reference to the "straight ahead" direction. If, for example, the strabismic eye is turned in by 10°,
then images formed on the fovea of that eye are interpreted as being located 10° from straight ahead. The fovea of the straight eye
no longer corresponds with the fovea of the turned eye but instead
with an area of the turned retina shifted by 10°.

This is how my friend Bruce Alvarez sees. He is a computer
scientist, a good ice skater, and a strabismic. His response to the
hole-in-the-hand experiment is different from what you probably observed if you tried it. He sees the hand in half of his visual
field and the tube of paper in the other half. Bruce does not
merge the input from the two eyes but instead reports what each
eye sees separately. Although this is not the response that most
people have, his view of the world, in this case, is actually the
more accurate one. After all, the tube of paper is to one side of
the free hand. Indeed, it's surprising to him that someone would
have a different response. Bruce's unusual way of seeing is called
anomalous correspondence, an adaptation that often takes
months or years of childhood strabismus to develop. It is a striking example of neuronal plasticity, a change in the way the brain
handles sensory input.

Dr. Frederick W. Brock, an optometrist practicing in the mid-
1900s, was an expert in treating strabismus and wrote many papers on the subject. He called patients with good acuity in both
eyes and anomalous correspondence "fully adapted strabismics."
Although a strabismic can suppress information in order to have
a single view of the world, a great deal of data is lost. But with

anomalous correspondence, a strabismic can effectively make simultaneous use of the information from both eyes. In fact, Dr. Brock described a cross-eyed truck driver who gave up driving his truck after his eyes were surgically aligned. Prior to surgery, he could look straight ahead at the road with one eye and look at a sign at the side of the road with the other. But once his eyes were straightened, he lost this useful panoramic view. He found it much more difficult to drive when he had to turn both eyes to read a road sign instead of reading the sign with one eye and keeping the other eye on the center of the road.

While investigating the ways his patients adapted to visual problems, Dr. Brock studied the work of neurologist Kurt Goldstein. In his book, *The Organism,* Goldstein stresses that a patient's symptoms are often a response to, or coping mechanism for dealing with, his or her disorder. With Goldstein's research in mind, Brock noted that a strabismic "speaks a different language." Eye misalignment hinders a person's ability to know where objects are located; as a result, strabismics develop different ways of seeing. Indeed, we cannot understand strabismus or any other visual disorder unless we consider how we use our vision on an everyday basis just to move from place to place.

Like cross-eyed individuals, people with many different deficits or disorders find surprising ways to adapt. Before becoming an astronaut, my husband was a physician specializing in physical medicine and rehabilitation. He took care of many patients with neuromuscular problems. He recalls one day as a new doctor when he was asked to examine a man with polio. Dan introduced himself to the man sitting on the examination table and proceeded to test the strength of the patient's leg muscles, quadriceps and hamstrings. He found them to be far weaker than normal and, after leaving the examining room, reported his results to the attending physician.

"Can the patient walk?" asked the senior doctor.

"I didn't test that," Dan said—but he assumed that the man could not.

So, Dan and the attending physician went back into the examining room and asked the patient to stand up and try to take some steps. To Dan's astonishment, the man got up off the table and confidently walked across the room. What's more, to Dan's unschooled eye, the patient's gait looked pretty normal. By any textbook description of locomotion, this patient should not have been able to walk—he simply didn't have the necessary strength in his quadriceps and hamstrings. But he had adapted; he had learned to recruit additional muscles in his legs because he needed to find a way to move in everyday life. Dan had been so focused on testing each individual muscle that he didn't ask the patient how he managed day to day. To Dan, this experience was particularly ironic because his own father had suffered from polio as a boy. His father's leg muscles were also weak, and Dan knew that, but until this incident, he had never wondered how his father had relearned to walk or climb stairs.

Adaptations, however, can come at quite a cost. No one knows this better than my friend Rachel Hochman. Rachel is a botanist, educator, and naturalist. She has accomplished all of this despite having been born premature and with a cataract, a cloudiness of the lens, of her right eye. When Rachel was a baby, she couldn't see clear images with her right eye, so she depended upon her left eye for vision. When she was four years old, her surgeon removed the clouded lens and replaced it with a contact lens, but by this time, her right eye's vision was seriously compromised. For the next four years, Rachel's left, or "good," eye was patched for part of each day in an attempt to strengthen the connections between the right eye and the brain. The visual acuity in her right eye never achieved normal levels, although it improved further with later surgery that included a lens implant. Unfortunately, Rachel had never used her two eyes together, so her brain still ignored much of the input coming from her right eye.

At age thirty-six, Rachel started optometric vision therapy with optometrist Dr. Hans Lessmann in Pittsburgh, Pennsylvania. Dr. Lessmann, an intense and compassionate doctor, has a keen interest in the relationship of vision to development, movement, and behavior. He noticed right away that Rachel's tendency to rely on her left eye for vision translated to the way she moved her body. During one therapy session, Dr. Lessmann tossed a ball to her and was flabbergasted by her reaction. Although Rachel was a rock climber and skilled kayaker, she was completely inept at catching a ball with two hands.

Catching a ball with both hands requires simultaneous mirror-image movements of the right and left side, but Rachel's left hand reached out too quickly while her right hand moved too slowly. Suspecting that her one-sided vision had disrupted her coordination, Dr. Lessmann assigned her an additional task—to move her right and left limbs in alternation, all to the beat of a metronome. Rachel overreacted with her left side and underreacted with her right. Her left side dominated the movements even though Rachel is right-handed. What's more, her hips were out of alignment with her right hip rotated toward the right side.

Dr. Lessmann now understood the source of the problem. Rachel used her left eye to see, so she turned her head—in fact her whole body—to the right in order to point her left eye straight ahead. Just as her left eye compensated for her right eye, her left arm and leg compensated for her right side. Rachel's adaptation to her one-sided way of seeing had distorted her posture and affected the way she moved. Her vision affected her entire body.

Like Rachel, my friend Tracy Gray suffered from a twisted posture associated with her vision. Tracy was a marathon runner, and her endurance and discipline kept her going despite a moderate torticollis, or an involuntary twisting of the neck and head to one side. Tracy found it almost impossible to straighten her head. When she wanted to look directly at people, she would cross her arms and bring her right arm up to her chin to keep her head straight. To drive, she would sometimes lean her head against the driver's-side window so she could steady her head's position.

For many months Tracy sought help from various clinicians to cope with the constant pulling of her head. Finally, an astute muscle therapist suggested that Tracy's problem was related to her vision and recommended that she consult Dr. Amiel Francke, a highly creative optometrist experienced in vision training. He told Tracy that she took in visual information with more effort than others. She had trouble coordinating her eyes—converging them for near viewing and diverging them for far. To compensate, Tracy turned her head so far to the left that her nose blocked the view from her left eye. With Dr. Francke's guidance and training procedures, Tracy learned to use her two eyes together, which led to a realignment of her posture and relief from the constant pulling of her head.

Fortunately, I did not adapt to my misaligned eyes by developing a distorted posture. Instead, I paid attention to the input from only one eye at a time. I switched rapidly between the two views, which made my world unstable or jittery, particularly when looking out in the distance. Not surprisingly, I was a pretty lousy driver. When Dan and I started dating, he was appalled by my jerky, start-and-stop style behind the wheel.

"Where are you looking when you drive?" he asked the first time he was my passenger.

"I'm looking straight ahead," I responded, irritated with him for making me talk while I was trying to concentrate on the road.

"How far in the distance are you looking?" he asked.

"I don't know. Maybe one or two car lengths ahead of me."

"That's what I thought," he said. "Try looking much further in the distance."

But looking in the distance was unnerving. I felt disoriented, unsure of my location in space. I felt like the car was drifting off the road. Frustrated, I gave up on Dan's suggestion and went back to my old driving habits.

Years later, Dan tried again, using a different tactic.

"You know what Kent told me today?" he asked, referring to Kent Rominger, the commander on Dan's second space shuttle mission on STS-96. Kent was a crackerjack pilot, who during STS-96 became the first pilot to dock the space shuttle with the International Space Station, a task a little like parking a car, traveling at Mach 25, in a parking space that is also moving.

"He said that the whole trick to piloting is to look in the distance. That way, you know what is coming up and can move the vehicle [typical astronaut jargon] accordingly."

Although Dan's remarks made sense to me, they didn't change the way I drove because I simply couldn't look comfortably more than a couple of car lengths away. I came up with a more passive solution: I avoided driving as much as possible.

My reluctance to look into the distance led to other, sometimes comical situations. About seven years ago, I was coteaching an eighty-student introductory biology class with one of my sharp-eyed colleagues. She asked me why I never acknowledged the students who raised their hands in the back of the large classroom. "What students?" I asked. With my glasses, I had 20/20 acuity in both eyes, but it took effort to keep my gaze steady while looking to the back of the lecture hall. Unconsciously, I had solved the problem by no longer looking there.

My colleague wisely took matters into her own hands. She positioned herself in the classroom behind all of the students. Whenever a hand popped up in the back of the class, she would wave her arms wildly to get my attention and then point in an exaggerated manner in the direction of the student with the question. This helped me to pay attention to all of the students, and for their part, they were nice enough to pretend that they saw nothing out of the ordinary.

Incidents like this made me feel like Mr. Magoo, the clumsy, inept cartoon character who bumped into everything because he was so nearsighted. But I could read an eye chart as well as the next person. My acuity could not explain my troubles. The ophthalmologist I had seen when I was forty had said that there was nothing wrong with my vision, and yet I still felt that I was missing something. As I stood in the front of my large class and noted how vigorously my colleague was waving her hands to get my attention, I realized just how much I had limited my own field of view. I decided that I should see another eye doctor, perhaps someone with a different approach.

I consulted Dr. Steven Markow, an optometrist whose office was located across the street from Mount Holyoke College. He referred me to Dr. Theresa Ruggiero, a developmental optometrist who practiced in a nearby town and did something called optometric vision therapy. The next day, I called Dr. Ruggiero's office and mentioned that I was almost forty-eight years old. Was I too old for vision therapy? Unfazed, the receptionist told me that they treated people from nine months to ninety years. "In terms of age, you are only halfway there," she said. Skeptical but hopeful, I made an appointment.

As I drove to Dr. Ruggiero's office for my first visit in November 2001, I wondered what she could do for my vision that other eye doctors hadn't already tried. After all, I had been the surgical patient of a world-class and well-known ophthalmic surgeon. Every eye doctor whom I visited after my surgeries had commented on how nicely my eyes had been aligned, how good my eyes looked. I remember well how embarrassing and humiliating it was to appear cross-eyed when I was younger, so I was very appreciative of looking normal. My surgeries were done so well that I had developed little scar tissue and could still move my eyes freely in all directions.

My doubts about seeing my new eye doctor didn't diminish when I walked into her optometry office. It was neat and organized, but it was modest and unassuming—there was no fancy medical center in sight. I was to learn, however, that the office's unpretentious appearance in no way reflected the extraordinary work that went on there.

Theresa Ruggiero had attended classes on sensory perception while majoring in psychology in college. She became fascinated by the role that experience plays in our perception—what we have seen greatly influences what we will see. As a research assistant for one of her professors, she loved conducting experiments on human visual perception. When she went back to her local optometrist for a routine eye exam during college, she told him about her perceptual studies and asked him about his profession. He suggested that she visit the offices of several optometrists who practiced vision therapy. Since optometric vision therapy emphasizes the connection between the eye and the brain, Theresa realized that a career as an optometrist specializing in vision therapy would allow her to put her academic ideas and interests into practice.

There are two main groups of eye-care professionals, ophthalmologists and optometrists. Ophthalmologists are medical doctors who spend four years in medical school learning about the body and diseases in general. During their postgraduate studies and residencies, they become skilled at treating disorders and diseases of the eye using pharmacology and surgery but not vision therapy. Optometrists, like Theresa Ruggiero, attend a four-year optometry school where emphasis is placed on the connection between the eye, vision, and the brain. Dr. Ruggiero chose optometry because of the unique field of optometric vision therapy—she wanted to work with patients who suffered from visual problems that could be helped by training the eyes and brain to see in new ways.

Optometric vision therapy has its roots in orthoptics, a group of techniques originally designed to straighten a strabismic's eyes. This therapy does not include popular self-help techniques such as the Bates or See Clearly methods. Instead, it is built upon orthoptic procedures developed in the late 1800s by the French ophthalmologist Louis Emile Javal, a social reformer who advocated for better conditions for the poor, improved treatment for the blind, and the adoption of the international language of Esperanto. In addition, he was the first person to measure eye movements during reading. Javal's interest in strabismus stemmed from concerns for his strabismic father and sister, and he developed therapeutic techniques to help them as an alternative to the crude surgery of the day, which he called *le massacres du muscles oculaire* (or massacre of the eye muscles). His techniques worked but often required long hours of practice to be effective.

Starting in the 1900s and continuing to the present day, ophthalmologists (especially in the United States) focus on surgery as a treatment for strabismus. As mentioned in chapter 1, laboratory experiments indicate that eye misalignment during a "critical period" in early life disrupts the development of binocular neurons. Many doctors interpreted these results to mean that normal binocular vision and stereopsis could develop only during these early years. As a result, ophthalmologists, in the late 1900s, began to operate on strabismic infants within the first year of life. Surgery on such young children has been partially successful in allowing for the development of stereovision. Since pediatric ophthalmologists now perform surgery on infants, they pay little attention to therapy as a treatment for strabismus. The babies are simply too young to participate in vision therapy procedures. Although ophthalmologists work with orthoptists, a group of professionals who prescribe some therapies, the procedures provided are only of the most basic kind. Instead, it was a small group of optometrists, working during the mid-1900s, who perfected and expanded Javal's initial orthoptic techniques.

From the 1950s through the 1970s, Frederick Brock, William Ludlam, and other optometrists successfully treated people with strabismus using vision training. In a study to determine the effectiveness of this training, 149 patients (none of whom had undergone any surgery) received treatment sessions twice a week for twelve weeks. After this time, 75 percent of the patients boasted straight eyes and stable binocular vision with stereopsis. When they were reexamined several years later after having had no further therapy, their eyes were still straight, and they still saw with stereopsis. Particularly intriguing was the fact that half of the patients who had developed strabismus in the first year of life gained stereovision, a result that conflicted with the interpretation of laboratory studies on animals. Several other investigators confirmed these important and groundbreaking studies. Yet, with the growing momentum behind the idea of the critical period and in favor of very early surgery, these reports were largely ignored.

Optometrists have always been at the forefront of lens development, designing, for example, the first contact lenses as well as low-vision devices (tools for reading and distance viewing used by people with severe vision loss). It is not surprising, then, that optometrists found ways to enhance therapies with the use of lenses and prisms. Since they are aware of the intimate connection between vision, posture, and movement, they incorporated training in visual motor skills into therapy procedures. In 1971, a subgroup of optometrists particularly interested in vision therapy founded the College of Optometrists in Vision Development (COVD) in order to standardize therapy protocols and develop a rigorous test for board certification in optometric vision therapy. Optometrists who pass the board-certification exam become fellows of COVD. At the time of this writing, only four hundred optometrists are fellows, while perhaps another five hundred are also well versed in vision therapy techniques.

While in optometry school, Dr. Ruggiero made sure to attend vision therapy clinics, particularly those of Dr. Israel Greenwald, an optometrist who had practiced with the late Frederick Brock. She received her board certification in optometric vision therapy and had been in practice for fifteen years when we first met. She treated children who struggled in school because of vision problems, people like me with binocular vision disorders, and individuals who had suffered from traumatic brain injuries and stroke.

I have been to eye doctors all of my life, and I thought I knew all about vision exams. Most eye doctors had checked my vision by testing one eye at a time, although I had occasionally been asked to follow a pencil or penlight as it was moved toward my face or to take a simple stereovision test. In Dr. Ruggiero's office, I performed a host of binocular vision exams that were new to me. When the tests were completed, Dr. Ruggiero looked over all the results, but before she described them to me, she asked me what I would like to do that I couldn't do at the time.

I told her that I wanted to be able to read for longer periods, to look comfortably in the distance, and especially to drive without feeling so anxious. I avoided going places or visiting friends if I had to drive to an unfamiliar location. I would come up with excuses because it sounded pathetic to say that I was afraid to drive. I had been in the car countless times when my husband was driving. I just couldn't understand how he sped up on entrance ramps to highways and merged with the oncoming traffic with such confidence and ease. I wanted to play tennis without my eyes tiring after twenty minutes. After I played tennis on an indoor court, I would experience a strange sensation: the whole world moved up and down when I walked. I wasn't looking for supersharp eyesight, but I was hoping for more comfortable vision for the routine activities of everyday life.

Dr. Ruggiero told me that my problems with eye fatigue, distance viewing, and driving stemmed from the fact that I was uncertain about where I was in space. I saw objects from two different directions, giving me a constantly shifting worldview. Her first and most important goal for me was to stabilize my gaze so that I could earn confidence in my own vision.

Dr. Ruggiero went on to explain that my eyes, though cosmetically straight, were still both horizontally and vertically misaligned. In addition to my horizontal strabismus, my right eye saw several degrees below my left eye. Other doctors had noted this mismatch but had chosen not to correct it. They may have assumed that I could not use my two eyes together, so there was no need to align my eyes any further than was cosmetically necessary. What they didn't realize, however, was that the vertical mismatch was creating visual havoc for me. As I unconsciously and rapidly switched my attention from one eye to the other, I was experiencing two significantly different views.

When Dr. Ruggiero explained all this to me, the Beatle's song "Nowhere Man" suddenly popped into my head. In the song's refrain, the Beatles tell of a man with no point of view who "knows not where he's going to." I had just learned that I did not have a single point of view; instead I had two! This situation had created

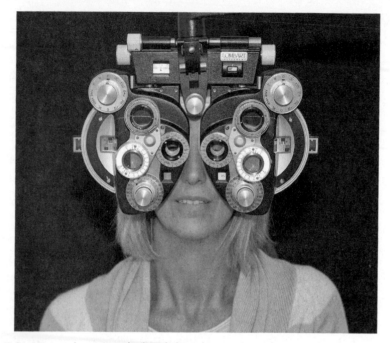

FIGURE 4.4: A patient behind the phoropter. (© Rosalie Winard)

subtle, but pervasive, problems with knowing precisely where I was in space.

Dr. Ruggiero placed a phoropter in front of my face. If you have ever been to an eye doctor, you've probably seen this un-wieldy device packed with lenses and knobs (see Figure 4.4). Using the phoropter, Dr. Ruggiero placed different lenses in front of my two eyes, including prisms, which can shift the entire vi-sual field of the eye up, down, right, or left. She put a prism in front of my right eye that shifted its view upward to be more in line with the view from my left eye, then asked me to read an eye chart displayed on a computer monitor twenty feet away.

"Does the eye chart look any different from how you'd normally see it?" Dr. Ruggiero asked.

"Yes, it does," I said. "It is easier to maintain my gaze on the letters. The letters don't jiggle." With the help of the prism, the act of seeing was so much easier. I was ready to walk out of her office wearing the whole phoropter in front of my face!

Dr. Ruggiero prescribed a new set of glasses with my usual corrections for nearsightedness but added a prism that would reduce the vertical mismatch between my eyes. Once my glasses arrived, Dr. Ruggiero told me, she would start me on a vision therapy program that would teach me how to coordinate my eyes and stabilize my gaze.

I left Dr. Ruggiero's office feeling hopeful about my vision for the first time. It was a tremendous relief to learn that there was an explanation for my unstable vision, that I had a legitimate, treatable complaint. Very few individuals with infantile strabismus discover what it is like to have stable, clear binocular vision with stereopsis, while most doctors have no personal experience with the disorder. There is a huge gap, then, between our technical knowledge of strabismus and the experience of seeing with misaligned eyes. Over the next year, I was to learn just how great this gap is. With therapy, my vision improved and my worldview transformed in ways that I could never have imagined.

5

FIXING MY GAZE

Perception is not something that happens to us, or in us.
It is something we do.

—Alva Noe, *Action in Perception*

It is Saturday morning, and I am in the supermarket about to pick out the best-tasting fruits and vegetables for my family. I am faced with a dilemma for I cannot take bites of the apples or sample the peaches. Instead, I must use every sense other than taste to predict how delicious the food will be. So, I approach the produce with the tenderness of a lover. My eyes move over the bell peppers as I take in their curvy shapes and varied colors, from dark green to bright red. I pick up a pepper and run my fingers over its smooth surface, checking for soft spots. I bring the pepper up to my nose to take in its sharp odor. I walk over to the display of avocadoes, again scanning the selection with my eyes. I pick up and squeeze an avocado to feel how much it pushes back, mentally making a note about whether I will be making the guacamole today or tomorrow. I bring the honeydew melon to my ear and shake it, listening for the rattle of the seeds inside. I cannot sample the fruit without moving first—moving my eyes to see, my hands to feel, and my nostrils to smell or moving the fruit to my ears to listen. We are often taught that we sense the world and then react. But our sensing and moving do not happen in sequence. We cannot perceive the world in any detail without moving at the same time. In fact, the act of planning the movement, which is largely unconscious, and the movement itself may sensitize our eyes, ears, and fingers. Perception and movement are intimately linked in a continuous two-way conversation.

In a series of extraordinary experiments, Paul Bach-y-Rita and his colleagues demonstrated the importance of self-directed movement in perceiving the world. They used a TV camera to translate the light/dark pattern of a visual scene onto an array of stimulators. Blind volunteers leaned their backs against this stimulator array and felt the tactile pattern of the visual scene. With training, the subjects reported that they perceived the pattern not as sensations on their skin but as images out there in space. They were able to use the TV camera system to recognize faces, accurately hit a rolling ball with a bat, and play the game rock-paper-scissors. This transformation happened only after the blind volunteers learned to control the movement of the TV camera. Only then could they use the camera to "see." Sensing the world is not a passive process but requires active exploration.

So, to learn to see better, I had to relearn how to move my eyes. This was the first step in my vision therapy. For my first vision therapy session, I was led through Dr. Ruggiero's suite, past all the familiar-looking rooms associated with an eye doctor's office, into what looked like a child's playroom. A tennis ball was suspended with a string from the ceiling. Red and green translucent plastic sheets were taped to a glass door. A giant pegboard displayed coiled loops of strings threaded with colorful wooden beads as well as a whole variety of red/green and Polaroid glasses. On the counter was a stereoscope and little wedge prisms attached to short posts. Charts with various letters and numbers were pasted to the wall, and boards were placed on the floor, apparently meant for patients to balance on. I was skeptical. This didn't seem like a place for serious, sophisticated work. And my concerns were not helped by the fact that I was sharing my vision therapy session with a five-year-old girl.

For the next year, I was coached in vision therapy procedures during forty-five-minute meetings once per week. Before each session, Dr. Ruggiero reviewed my progress from the previous week and prescribed the current week's therapy. After each meeting, her vision therapists sent me home with instructions to practice different procedures for twenty minutes each day. During one meeting, vision therapist Michelle Dilts gave me four sheets of paper, each with a large number printed on it. She told me to tape one piece of paper to each of the four corners of a wall at home. With a patch over one eye, I was supposed to look at one and then the next number. I was to have someone call out the numbers in random order, look quickly at the paper with that number, hold my gaze for a few seconds, and upon a command from my helper, shift my gaze to a new number in a different corner of the room. "You've got to be kidding me," I thought to myself. "This is an exercise for a child."

Michelle must have read my mind because she asked the little girl who was in session with me to demonstrate. This small, bespectacled child could move her eyes and hold her gaze much better than I could. After this demonstration, Michelle didn't need to say anything more. Feeling a bit sheepish, I took the number sheets, brought them home, taped them to my kitchen wall, and practiced with them every day.

To hold your gaze steady seems like a simple task, one that should require no effort at all. In fact, this important skill involves a large number of brain regions. If the image of an object on your retina does not move at all, it fades away. So, when you look steadily at an object, you must continually refresh the image by moving your eyes subtly, for instance, by slowly sweeping them across the object and making small, quick jerks.

Not only do we orient ourselves in the world with our gaze, but we also use it to communicate with others. A nursing infant gazes steadily into her mother's face, and her mother gazes back. Lovers gaze into each other's eyes. If we want to intimidate someone, we tend to stare directly and steadily at that person. We fix our gaze, avert our gaze, lock our gaze, and gaze into the distance. A simple shift in gaze can signal a change in mood or thought. Our gaze has meaning because it is a purposeful directing of our eyes.

When I look through old family photos, I'm hard-pressed to find a baby picture of me that shows my eyes actually gazing at the camera. My eyes are always looking someplace else. All through childhood, my sister had complained that my gaze wandered when I talked to her, making it seem like I wasn't paying attention. "Where are you looking?" she would ask as I stared off into space. Even after my surgeries, I couldn't maintain a steady gaze for many seconds. Sometimes, the scene would become grainy, and I'd shake my head to get it back.

Up until I started vision therapy, I found it particularly difficult to hold my gaze steady while looking down. So, I looked down by lowering my whole head. And each morning, I opened my eyes to a drifting world. My whole surroundings would slip to the right, then jerk back to the left, only to drift to the right once more. Sometimes, I would lie in bed and just watch the world slip by for a few cycles before stopping the drifting view with a shake of my head. I was so used to these instabilities that I was not too concerned about them and didn't report them to an eye doctor. But Dr. Ruggiero noted my abnormal eye movements in her first exam. If she covered one of my eyes, the uncovered eye would look directly at the target she asked me to view, then start to drift in toward my nose, zing out to refixate on the target, and then start to drift noseward again. This abnormal movement, called latent nystagmus, is often seen in people who have been cross-eyed since infancy.

With several vision therapy procedures, such as the "four-corners" exercise, I slowly learned how to hold my gaze. I realized that I had never really read the words on street signs because I didn't let my eyes rest on a small target even for a moment as doing so took a great deal of effort. Instead, I kept my eyes moving, interpreting the sign's message (often incorrectly) from context and by guessing at the identity of the largest letters. So, now, when out walking, I practice looking at street signs from incrementally greater distances, aiming my eyes at, and holding them steady on, each individual letter.

I practiced some basic eye movements too, like saccades, which are rapid movements of both eyes in the same direction. The word "saccade" derives from the French and means "to jump." In fact, it was Louis Emile Javal, the man who developed orthoptics, who coined the term. Saccades are the fastest type of movement that we can make, occurring in as little as one-fiftieth of a second. We scan our surroundings by making several saccades per second, although we are rarely aware of doing so. When we turn toward a new object and pick it up, we first make a saccade, then a head turn toward the object, followed by reaching movements. As you read this book and come to the end of a line, your eyes make a rapid saccade leftward to fixate the first word of the next line.

During one early vision therapy session, Dr. Ruggiero introduced me to a new procedure involving a ball hanging from the ceiling by a string. She set the ball into gentle motion and asked me to follow it with only my eyes, a task that requires slow eye-tracking movements called smooth pursuits. After a few moments, Dr. Ruggiero reminded me to move only my eyes and not my head. "I'm not moving my head," I insisted. So, Dr. Ruggiero

handed me a book and instructed me to balance it on top of my
head. She set the ball in motion once again, and while I followed
its movements, presumably with my eyes only, the book fell to
the floor. I tried again, but the same thing happened. So, I learned
that I did not move my eyes independently of my head. I took the
ball home, strung it up to my kitchen ceiling, put a book on my
head, and practiced.

Smooth pursuit movements of the eyes are often abnormal in
people who have been cross-eyed since infancy, and this deficit
may result from poor development of stereovision. Poor eye movements probably explain why I could never see the ball in rapid play
when I sat midway back in the stands during Major League base-
ball games. I saw the ball if it was tossed casually from player to
player but not when it was in fast motion. Instead, I saw the
pitcher wind up, the batter swing, and the fielders react all in one
elaborate pantomime.

Once I realized how important gaze holding and eye movements were to good vision, I performed the gaze holding, ball
tracking, and other therapy procedures regularly. I practiced these
tasks while standing still, walking, and even bouncing on a
minitrampoline. And then, one day after about a year of vision
therapy, I woke up to a stable world.

My view was rock solid from the moment I opened my eyes.
I lay in bed waiting for the world to start drifting by, but it did
not. When I looked down, the world didn't appear to shake back
and forth. Now, when I'm done playing tennis with a friend, I no
longer experience the sensation of the world moving when I walk
off the court. To my delight, on a recent trip to Fenway Park, I
watched a Red Sox game from midway back in the stands and
saw the ball in play most of the time.

After I learned to fix my gaze, I not only saw differently but
looked different as well. I could gaze comfortably without having

to hold my eyes wide open. My husband is thrilled that my vision has improved but tells me that he misses my old saucer-eyed look. He found it very appealing.

To see comfortably, I needed to develop a stable gaze not just while standing but also while moving. Seeing clearly while moving is critical to good vision, but the typical eye exam rarely tests this ability. In a standard exam, you sit quietly in a chair while viewing with one eye an unmoving eye chart positioned twenty feet away. This test measures the smallest detail you can see with one eye at that distance, but there's more to vision than that. During the course of an ordinary day, you look at objects located both far and near, and some of them are moving while you're in motion. You take in the world by merging the information from two eyes. What's more, you combine information from the eyes and other senses into a perceptual whole and direct your movements according to what you sense. None of this is tested with the standard eye chart. To really examine how well you see, an eye doctor needs to know how well you perform the tasks of everyday life, such as reading a book, driving a car, catching a ball, or even looking about while walking down the street.

When I worked at NASA's Johnson Space Center in the 1990s, several years before my own vision therapy, we tested people's dynamic visual acuity, that is, how clearly they saw while moving. When some astronauts first return from space, they complain that the world appears to move when they turn their heads or walk, an effect that could cause serious problems if they have to perform emergency procedures right after landing. Since their vision seems less stable when they move, we wanted to compare their dynamic visual acuity upon returning from space to that of "normal," earthbound people. We asked earthbound people and returning astronauts to read an eye chart while standing still or while walking on a treadmill. Most earthbound people could read the eye chart just as well while standing or walking. I had normal 20/20 visual acuity while standing, but my acuity dropped dramatically while I was walking at four miles per hour. No wonder I could never read labels on packages while walking down grocery store aisles the way other people could.

Similarly, some astronauts experience poor dynamic visual acuity when they first return from space. This temporary reduction in vision results from disrupted coordination of eye, head, and body movements. You can experience what the returning astronaut may see by taking a video camera and filming the scene around you as you walk along. When you look at the tape, you'll see that the images in the video bob up and down with your own movements. Yet, when we walk and look at the world through our eyes, the world is pretty stable. How do we accomplish this?

The following simple experiment, a favorite with my students, illustrates in part how we keep our gaze stable while moving: Look in the mirror and direct your gaze straight ahead. Now, slowly move your head left and right as if you are shaking your head to indicate "no." Keep looking straight ahead. You'll notice that when your head moves right, your eyes move left and vice versa. The faster you move your head (within limits), the faster your eyes move. Move your head up, and your eyes move down; move your head down, and your eyes move up. This all happens involuntarily. Since your eyes move in a direction opposite to that of head movement, you can keep your gaze pointing straight ahead.

Compensatory eye movements like this are mediated by a combination of the vestibulo-ocular reflex and smooth pursuits. Inside your ear, along with the cochlea, a structure for hearing, is a set of other structures that make up your vestibular organs. These structures, called the otolith organs and semicircular canals, are exquisitely sensitive to the movement of your head. When these structures sense head movement, they send this information to neurons in your brain that stimulate, in turn, different eye muscles, causing your eyes to move in a direction opposite to your head. This is the vestibulo-ocular reflex.

The reflex alone does not provide enough eye movement to cancel out the effects of the head movement completely, so your eyes also make an additional pursuit movement in the opposite direction of the head. As a result, the world does not appear to move as you move. The average video camera does not have a built-in vestibulo-ocular reflex, so when it moves up and down as you walk, it takes frames from many different angles. Watching the video gives you a sense of how the world would appear when walking if you did not have compensatory eye movements.

Not surprisingly, people who suffer injury to their vestibular system have trouble keeping a stable gaze as they walk. The astronauts who adapt most easily to spaceflight are the "head lockers," those who move their eyes, head, and body in unison when they are first afloat in space. In this way, they minimize the need for compensatory eye movements, depend less on their confused vestibular system, and are able to maintain a steady gaze.

People with strabismus and other vision disorders may also have trouble stabilizing their gaze and seeing clearly while moving. Eric Woznysmith, a strabismic, told me that before his vision therapy, he felt like he was "walking through life on slippery stones," and I knew just what he meant. Before I met Dr. Ruggiero, I unconsciously developed strategies to get around this troubling problem. When I went out jogging, my friends would tell me that they would pass me by, hanging out of their cars, shouting, and waving like mad, but I never saw them. I didn't move my eyes or head to the right or left but kept them pointing straight ahead in order to keep my view as stable as possible.

Following my vision therapy, my ability to see clearly while moving improved dramatically. Not only could I read labels and grab items off grocery store shelves without having to stop, but I could catch a Frisbee while turning in mid-jump. Now, I could drive our car up to an intersection, look both ways, determine if the road was clear, and then turn the car onto a new street in one smooth, continuous motion. Needless to say, my tennis game improved.

———

Optometrist A. M. Skeffington is often considered the father of vision therapy. In the early 1900s, he traveled around the country giving lectures, emphasizing that good vision means much more than 20/20 eyesight. Instead, vision is a learned behavior that can be improved with training. While walking by a train yard one day, Skeffington observed a girl playing on the railroad tracks. Being a friendly and fun-loving sort, he decided to join her and started to walk toward her while balancing on the tracks. As they approached each other, he couldn't help but notice her eyes. They appeared straight. However, once the girl hopped off the tracks and no longer needed to pay attention to her balance, her eyes crossed. Balancing and aligning her body correlated with balancing and aligning her eyes.

Observations like this one led optometrists Amiel Francke and Robert Kraskin to design balance boards in the mid-1900s to use in vision therapy. One type of balance board is a twelve-by-eighteen-inch piece of plywood centered on top of a foot-long two-by-four. To keep the board from rocking as you stand on it, you must distribute your weight equally. The board wakes up your vestibular system and reminds you to stay balanced and aligned.

Eliza Cole, a strabismic woman in her thirties, came up with a special use for her vestibulo-ocular connection. When Eliza started vision therapy, she began to pay attention to the input from both eyes and initially experienced double vision. This problem resolved as, with more therapy, she got better at aligning her eyes. For Eliza, her double vision was worse when she sat still. As a psychotherapist, this definitely caused problems for her at work. While she sat quietly, listening to her patients, they appeared to her to have two heads. Now, she wondered, who has a firmer grasp of reality, the therapist or the clients? Eliza had already discovered that movement made her double vision go away, but she couldn't get up and move about during these therapy sessions. So, she compensated by treating herself to a swivel chair and slowly rotated the chair to and fro while talking with her patients. The movement was subtle enough not to bother the patients but active enough to stimulate her vestibulo-ocular system.

Sharp central vision helps us to see acutely and to manipulate objects. However, to have a good sense of our surroundings, to know where we are in space, we need more than sharp sight. We need good peripheral vision. Keen peripheral vision is a skill we all appreciate when navigating, driving, or playing soccer, and it's also a skill that truly distinguishes the good from the exceptional athlete. When former senator Bill Bradley was a college basketball star, he was known for his perfect shots. If he found no teammate to whom he could pass the ball, then, with his back to the backboard, he tossed the ball over his shoulder right into the basket. When asked how he did this, he said that when you play on a basketball court long enough, "you develop a sense of where you are." Wayne Gretsky, the hockey star, and Larry Bird, the basketball champion, were also famous for their "field sense," a highly developed spatial sense that translated into extremely accurate movements. We need to use information from our entire visual field to develop a good spatial sense. We need to look far and wide.

I would never have made it as a basketball star because I was oblivious to most of my surroundings. Like many people with visual difficulties, I paid attention to only a portion of my visual field, in my case, to that which was up close, front, and center. My visual impression of a walk down a busy street used to be more like a series of discrete snapshots as opposed to a continuous flow. I was easily disoriented and constantly lost.

When I took my job at Mount Holyoke College in 1992, I dreaded driving in the neighboring town. The traffic lights in Holyoke are located on poles on the sides of the street instead of being suspended overhead in the middle of the street. I looked only straight ahead, so I tended to miss the traffic lights and careen right through the intersections. I should have been able to find the city hospital by following the hospital "H" signs, but those too are located on the sides of the road, so I missed them as well and got lost.

In *Living in a World Transformed,* Hubert Dolezal examines what it is like to see without peripheral vision by wearing a set of tubes over his eyes for one week. The tubes allowed him to see only 12° of the visual field as opposed to the normal 180°. His descriptions of what he saw sounded eerily familiar to me. While wearing the tubes, he felt easily disoriented, especially in unfamiliar environments, and was surprised when people or objects seemed to appear out of nowhere. He bumped into doorframes and other obstacles.

Dolezal also reported that he was totally unable to follow the action in movies. Perhaps my inability to take in the whole movie screen at once explained why I had a hard time following the plots in many films. I tended to nod off when the action got complicated, like when the Mir space station blew up in the movie *Armageddon.* The faster the action in a movie, the quicker I started to snore, much to the amusement of my children.

Even people with normal vision have different visual styles. As I became aware of the way I used my own vision, I started to divide people into "near lookers" and "far lookers." I'm still a near looker, though my vision is now more balanced. I am very cognizant of the space immediately in front of me, and I like it ordered and uncluttered. One of my colleagues, Tara Fitzpatrick, has the opposite style. Tara is walleyed and a "far looker." She finds it relaxing to look into the distance. As a result, she is an excellent navigator; she always knows which direction is north. But she has a hard time judging objects a few feet from her and is constantly tripping over curbs.

My husband, who has normal vision, is also a far looker. Dan loves to play basketball, tennis, and laser tag, all of which require a keen awareness of his surroundings. He is a relaxed and skilled driver and pilot. When working at NASA, Dan loved flying the acrobatic T-38 jets in formation. Wingtip to wingtip, the jets flew only four feet apart. He tells me that the key to driving on a highway is to "reduce your delta V to zero." Translated from engineering lingo, this means that you should match your speed with the speed of the cars around you, a task that requires good peripheral vision.

Dan can put up with a remarkable amount of clutter around the house. I used to take great offense at this since he knew that I liked things neat. It took me a long time to realize that he simply didn't notice the mess. Oh, he could see the clutter; his visual acuity is superb. But his attention was focused further away.

Dan also has excellent stereopsis, and as I will describe in the next chapters, stereopsis provides us with the ability to see the volume of space between objects and to see each object as occupying its own space. Without it, the world appears more contracted, and objects in depth seem piled up, one on top of the other. I recently interviewed a woman who lost vision in one eye as a result of a car accident. She now insists that her house and desk stay neat and organized, each object in its own place. A psychologist told her that her desire for tidiness might be borderline obsessive, but the psychologist hasn't experienced what it's like to see with only one eye. I believe that the woman's desire to organize is a way to adapt to her visual losses.

To teach me to make better use of my peripheral vision, Dr. Ruggiero had me play the "wall game," which involves the Wayne Saccadic Fixator, a device built by Harry Wayne of Wayne Engineering (Figure 5.1). The Wayne Saccadic Fixator contains several concentric circles of lights. One light goes on at a time, and you must hit it in order to turn it off and cause another light to turn on. The goal of course is to hit as many lights as possible in a certain amount of time. The game effectively builds rapid eye-hand coordination and is certainly popular among the patients.

Most people who play the game think that the best strategy is to continually and rapidly scan the board of lights in order to see when a light turns on. However, your perception is actually inhibited during a rapid scanning saccade. If this were not the case, the world would seem to fly by every time you quickly changed your gaze. Instead, the best strategy is to "look soft," to let your gaze expand to fill up the whole board, which enables you to find the lighted target more quickly.

A few years ago, I went to the office of optometrist Dr. Paul Harris to ask him more about his work with vision therapy. During my visit, a strong, fit young man who was a former football player came to deliver some equipment. He saw patients playing the wall game and wanted to give it a try. After the young man played, Dr. Harris casually suggested that his vision therapist, Liz St. Ours, take a turn. Liz St. Ours is a tiny woman, literally half the size of her opponent. She stood in front of the board and then with rapid moves, very much like Jackie Chan doing karate chops, she doubled

FIGURE 5.1: Heather Fitzpatrick playing the "wall game" with the Wayne Saccadic Fixator. (© Rosalie Winard)

the young man's score. The football star was flabbergasted and tried again, but he couldn't catch Liz's score. What was her secret? She "looked soft" and used both her central and peripheral vision.

At home, I still practice some procedures to promote this peripheral awareness. I found the old magnetic letters that had once decorated our refrigerator when my kids were first learning to spell. With my eyes closed, I spread these letters out in a random pattern around a central point. Then, while keeping my eyes on this central point, I retrieve the letters in alphabetical order. I have to look softly and make myself aware of my peripheral visual fields to do this.

Similarly, during our summers in Woods Hole, Massachusetts, I often bicycle into Falmouth along a bike path, a narrow asphalt road with a central yellow line. I try to keep the bike rolling right on the line. On my first attempt, I tried this by looking directly at the yellow line in front of me and found that the bike kept wobbling across it. So, I asked Dan what he would do, and he reminded me to look softly in the distance and take in the whole view. On my next journey, almost magically and with less effort, the bike rolled right along the yellow line.

I find the looking-soft technique useful in all sorts of places. When my father developed a tremor in his left arm, he had to give up violin playing. When I visit him now, we sit together and listen to recordings of Nathan Milstein playing Bach violin sonatas. As we listen, we follow along with the musical score. Milstein played with incredible speed and fluidity; the sixty-fourth notes just speed by. In the past, I would have tried to follow the music by looking at each measure separately and lost my place in the score. But now I take in the whole contour of the musical line. I can follow the musical landscape easily, with both my eyes and ears, and this gives me a new and special pleasure.

As I continued my vision therapy and became increasingly aware of my peripheral vision, I was able to tap into a phenomenon called optic flow. When you move forward, objects to the side of you appear to move backward. This optic flow is fastest for objects oriented at 90° to your movement, and the closer the objects are to you, the faster they appear to move. There is no optic flow for objects located directly in front of you: you know that you're getting close to them because they loom larger as you approach. Cinematographers and video game designers have figured out how to create illusions of motion on flat screens by simulating optic flow. Car-chase scenes in movies in which you see the surroundings pass by from the driver's-eye view make great use of optic flow. As my peripheral awareness improved, I began to use optic flow to gauge and, better yet, control the speed of my car. I felt myself moving smoothly in a large, stable, and stationary world. I started to relax. There are days when I even like driving.

Once I learned to move my eyes more efficiently, I was ready for an even greater challenge—learning to take in information from both eyes at the same time. To do this, I had to break a life-long habit of looking with one eye and suppressing input from the other. Since I had spent the last twenty years giving lectures to my students on visual development and critical periods, I told Dr. Ruggiero that I didn't think this was possible. My brain, I believed, was permanently wired for monocular vision.

But Dr. Ruggiero thought differently. She pointed out that it actually takes more brain activity to look through one eye and suppress the other than it does to look through both eyes. Suppression allowed me to have a single view of the world, but it is an active, demanding, and less efficient process.

When Dr. Ruggiero described suppression to me in this way, I was reminded of a lecture I gave in my neurobiology class about the way we move our fingers. I'd ask my students to roll their hand into a fist and then open their hand by extending their four fingers upward and their thumb downward. Then, I told them to close their hand into a fist again and lift only the index finger without moving the other fingers or thumb. Which action, I asked, takes more neuronal activity, opening the whole hand or lifting just one finger? Certainly, more fingers are moved and more work is done while opening the whole hand, but it takes more neuronal input to lift only one finger. To lift just one finger from a closed fist, your nervous system activates the circuitry that opens the whole hand. Then it has to inhibit the neurons in the circuit that control movement of most of the fingers and thumb. Only the neurons that control movement of the index finger are not suppressed.

Just as it takes more neurons to lift one finger than to open the whole hand, it takes more neuronal activity to suppress input from one eye than to see with two eyes. As I will describe in chapter 8, even a strabismic child starts out with a binocular visual system. It takes added neuronal input to disrupt this binocularity. Not surprisingly, then, the degree of suppression varies with viewing conditions. Suppression is strongest under natural, daytime viewing conditions, under conditions when a strabismic most needs a single view of the world.

Dr. Ruggiero gave me a set of red and green panels just like the ones I had noticed taped to the windows in her therapy room. Each translucent panel was 5.5 inches wide by 8.5 inches tall. I was instructed to tape these panels to a window with the red panel on the left, in line with and just touching the green panel. I was to look at the panels while wearing a pair of red/green lenses over my regular glasses. Under these conditions, I could see the red panel only with the eye behind the red lens. The other eye would see the red panel as black. Similarly, I could see the green panel only with the eye behind the green lens. So, to see both the red and green panels at the same time, I had to be aware of what both eyes were seeing.

My husband came to pick me up from Dr. Ruggiero's office on the day I took home the red and green panels. While Dan was driving home, I put on the red/green lenses and held the panels very close to my eyes. I could see the color of both panels. "Hey," I said excitedly to Dan, "I think I see simultaneously with both eyes now."

However, things were not that simple. When I got home, I taped the panels to a window as I had been told to do. Then I walked some distance away, and one panel, or part of one, turned black. After a moment's thought, I realized what was happening. As I walked away, the panels were taking up less and less room in my visual field, so I had to aim my eyes more precisely to see the color of both panels at the same time. Otherwise, I would suppress

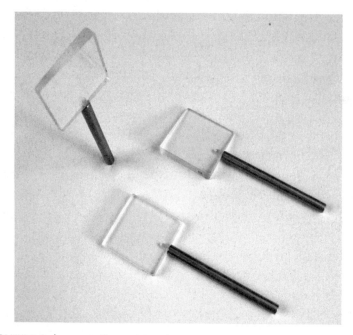

FIGURE 5.2: Loose prisms. (Photo by James Gehrt)

one eye's image, causing one panel to turn black. I had to practice aiming my eyes so that I saw all the color in both panels while
standing at different distances or while walking back and forth.

After a few weeks of daily practice, I had mastered the task to
a distance of eleven feet. I thought I was done with the red and
green panels. But Laurie Sadowski, Dr. Ruggiero's therapist, told
me that it was time to make the task more difficult, a process that
she called loading. She gave me a set of loose prisms of increasing
strength to hold over one of my eyes (Figure 5.2).

These prisms bent the light, which essentially shifts one eye's
view in the horizontal direction. While looking through the prism,
I had to realign my eyes slightly in order to see the color in all regions of both panels. I had to practice seeing both panels with

and without prisms at different viewing distances. So, I was back to practicing with the red and green panels for a few more weeks. Loading the task made this ability not only more automatic but also more fluid and more natural—as if I had learned it as a six-month-old.

If I turned on the input from both eyes, I asked Dr. Ruggiero, wouldn't I see double? Isn't this why I suppressed in the first place? This concern is one of the reasons why it is so important to undertake vision therapy with an optometrist highly trained in binocular vision. To prevent this situation, Dr. Ruggiero gave me several procedures that taught me how to aim both eyes precisely at the same place in space at the same time. As I mastered this skill, something remarkable happened. I learned to fuse images from the two eyes and achieved what I thought was impossible: I began to see in stereoscopic depth.

6

THE SPACE BETWEEN

It must be repeated here that, before stereopsis is actually experienced by the patient, there is nothing one can do or say which will adequately explain to him the actual sensation experienced. . . . Once the patient has experienced this new sensation, he is only too anxious to use it again and again until it is surely and definitely established.

—Frederick W. Brock

For most people, seeing in stereoscopic depth happens effortlessly. When we look at an object, our brain automatically compares the images seen by the right and left eyes. If the position of the image on the right fovea is a little different from that of the image on the left fovea, the brain commands the eyes to turn in or out to minimize this disparity. When I looked at something before I acquired stereovision, the object cast an image on the fovea of one retina but on a noncorresponding, nonfoveal point of the other retina. The disparity between these two images was too great for me to automatically make corrective convergence or divergence eye movements. I needed to learn very consciously how to aim my two eyes. To do this, I needed feedback that showed me where my eyes were pointing in space so that I could redirect them. Fortunately, Dr. Ruggiero had a solution, a simple device invented by optometrist Frederick W. Brock.

Dr. Brock was born in Switzerland in 1899 and came to the United States in 1921 to attend the Columbia School of Optometry. He was an inventive and playful man. When, one summer, he found himself without an outdoor grill, he simply fashioned one out of a garbage can lid. When his daughter went to summer camp, he sent her letters peppered with little drawings and visual puns. His lifetime passion was helping people with their vision, and in the mid-1900s, he helped many strabismics learn to see in stereo. Not content to test merely what his patients could not see, he performed many thoughtful experiments to determine what and how his patients did see. From these investigations, and with his typical resourcefulness, he built some of his own vision aids out of everyday items. In so doing, he taught us that you can learn a lot from a simple piece of string.

When I learned to use the "Brock string," I received the feedback that I needed to know where my eyes were pointing and then to redirect them so that they were aiming simultaneously at the same point in space. During one vision therapy session, Laurie took a five-foot-long piece of ordinary string on which were threaded three brightly colored beads. She attached one end of the string to a doorknob and gave the other end to me. She told me to walk backward until the string was taut and then to put the free end of the string up to the bridge of my nose (Figure 6.1). She slid one of the beads along the string so it was only a few inches from my face. Then she asked me to look at the bead and tell her how many strings I saw.

I thought that Laurie had asked a pretty ridiculous question. I looked down the string at the bead, saw one string, and told her so. Laurie nodded and then, while gently moving the string up and down, asked me to look again at the bead and tell her what I saw. To my surprise, I momentarily saw two strings emerging in front of the bead and two strings receding from the bead as in Figure 6.2.

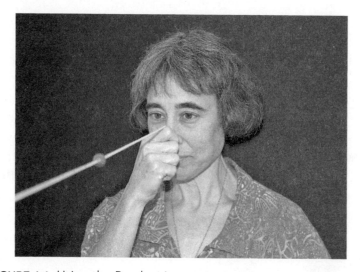

FIGURE 6.1: Using the Brock string. (© Rosalie Winard)

Why did I see four images of the string, two in front and two behind the bead, when only one string existed? To help me understand, Laurie drew the picture shown in Figure 6.3.

In this picture, the two foveas are pointing at the same place in space, the fixation point, which in my case was the bead on the string. The image of the bead fell on corresponding points on my two retinas, and my brain fused the bead images from the right

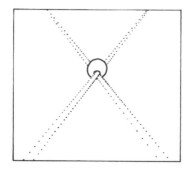

FIGURE 6.2. (© Julia Wagner)

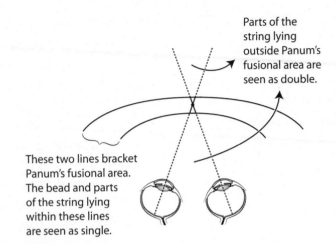

Parts of the string lying outside Panum's fusional area are seen as double.

These two lines bracket Panum's fusional area. The bead and parts of the string lying within these lines are seen as single.

FIGURE 6.3: Panum's fusional area. Each dotted line represents the line of sight for each eye. (© Margaret C. Nelson)

and left eye into one. There is a small area right in front of and behind the fixation point where the images fall on almost corresponding retinal points. This area is called Panum's fusional area, and I could also fuse the images of those parts of the string that fell within this area. All other parts of the string fell outside Panum's fusional area and were seen as double. So, I saw four images of the string, two in front of and two behind the bead.

This was fantastic feedback. Now I knew if I was suppressing the input from one eye because then I would see only one image of the string in front of and behind the bead. I could tell when my eyes were both aimed at the bead because then the string images would appear to emerge symmetrically from its center. With the Brock string, I could tell where in space my two eyes were looking.

Initially, I could "get" the four string images only when I brought the bead very close to my nose. If I looked at a more distant bead, I would lose one of the string images behind the bead.

Getting the second string image into consciousness was a strange experience. I would sometimes "feel" that the second string image had just been there, that I could almost reach out and touch it. If, at that moment, I made a slight eye movement, I might get the second string image to appear. This was incredible, a way of controlling what I consciously saw.

At other times, I would see all four string images, but they would meet not at the bead but in front of it. Then I knew I was aiming my eyes in front of the bead. I would touch the bead with my fingers so that my arm movements could help me to judge how far away the bead was and thus to direct my eyes accordingly.

Once I developed a range of distances over which I could properly aim both eyes at a bead, Laurie started me on a task involving two beads. She slid one bead close to my face and the other two feet away. She told me to look with both eyes first at the close bead, then at the further one, then back at the close one, and so on. She was asking me to converge my eyes to see the close bead and diverge my eyes to see the distant one. With a determined and conscious effort, I could turn in my eyes to point them at the closer bead and turn out my eyes to aim them at the more distant one. I knew that I had performed the procedure correctly if I could see the four string images emanating from the fixated bead. But, above all, *I could feel my eyes moving as a team! I could feel my eyes converge and diverge!*

We all have epiphanies, moments when something so simple, yet so evasive, hits us over the head. The Brock string brought me such an epiphany. I learned that I was able to coordinate my eyes for normal binocular viewing. I just needed the feedback to know where my eyes were looking. I thought initially that this new-found skill applied only to the task at hand—but I was wrong.

Movement alone enhances our perception. Dr. Brock emphasized this concept when he wrote about the importance of designing tasks for strabismics that they could complete only if they actively moved their eyes together, tasks beyond their "comfort zone" but not their capabilities. Most importantly, he realized that stable, clear binocular vision and stereopsis could be achieved only if the strabismic actively positioned his or her eyes, or made what Brock called a "fusion effort." I made my first fusion efforts when I aimed my two eyes together at the beads on the Brock string. What happened next amazed me, but Dr. Brock, had he been there, would have nodded knowingly and smiled.

———————

The sun was setting as I left Dr. Ruggiero's office after the long session with the Brock string. I got into my car, sat down in the driver's seat, placed the key in the ignition, and glanced at the steering wheel. It was an ordinary steering wheel against an ordinary dashboard, but it took on a whole new dimension that day. The steering wheel was floating in its own space, with a palpable volume of empty space between the wheel and the dashboard. Curious and excited, I closed one eye and the position of the steering wheel looked "normal" again; that is, it lay flat just in front of the dashboard. I reopened the closed eye, and the steering wheel floated before me.

I wondered if I was seeing in stereoscopic depth but reminded myself that this should not be possible. It was one day after my forty-eighth birthday. I was more than forty years beyond the critical period for the development of stereovision. Since the sun was low in the sky and shining light into the car at an odd angle, I told myself that the fading light must have created this unusual illusion.

The next morning, I practiced the Brock string procedure for ten minutes before getting in the car to begin my routine drive to work. As I looked up to adjust the rearview mirror, the mirror

popped out at me, floating in front of the windshield. I was trans-
fixed. Throughout the day, my stereovision would emerge—
intermittently, fleetingly, unexpectedly—bringing me moments
of absolute wonder and delight. The most ordinary objects looked
so beautiful. A large sink faucet reached out toward me, and I
thought I had never seen such a lovely arc as the arc of the faucet.
The grape in my lunchtime salad was rounder and more solid
than any grape I had ever seen before. I could see, not just infer,
the volume of space between tree limbs, and I loved looking at,
and even immersing myself in, those inviting pockets of space.

This new way of seeing was confusing as well, as I soon discovered while walking to my office one day. On the ground floor
of the biology department where I work, we have a skeleton of a
large horse displayed along with skeletons of smaller creatures and
cases full of stuffed birds. I have passed these rather spooky dis-
plays every morning for the past ten years, giving them little attention or thought. But just eight days after I began to see in
stereo depth, I was walking to my office and happened to glance
directly at the horse's head. The horse's skull, with its large teeth
and two empty eye sockets, loomed so far out in front of its body
that I thought it was moving toward me. I jumped backward and
cried out. Fortunately, no one was around to witness my panic.

This sense of objects projecting straight toward me was novel.
Objects had always appeared a little to one side, depending upon
whether I was paying attention to the input from my right or left
eye. Now that I was developing stereovision, I saw the horse's head
projecting straight toward me in the direction of a virtual "cyclopean
eye." The same was true for my view of car bumpers, open doors,
light fixtures, tree limbs, and outside corners of large buildings.

Shortly after my vision began to change, I went out for a daily
jog. My attention was drawn to the leaves on my neighbor's bushes.
Each leaf, I realized, occupies or captures its own space. Two leaves

cannot occupy the same space. In fact, two things can't be in the same place at the same time.

How profound and yet how obvious! I had always known that two things can't be in the same place at the same time because, for example, I could not force two pegs into one peg-sized hole. But now I had *seen* this. To see something for the first time that you know to be true is a deeply gratifying experience.

During this period, I reread *The Man Who Mistook His Wife for a Hat*. In a story entitled "Hands," Oliver Sacks beautifully illustrates the role of action in perception. He tells the true tale of Madeleine J., a woman who entered St. Benedict's Hospital in New York City when she was sixty years old. She had been blind since birth and had cerebral palsy, which severely restricted her movements. All of her life, she was dressed, fed, and cared for by others, so she never made use of her hands. They are "useless, god-forsaken lumps of dough," she told Dr. Sacks. Yet, when Dr. Sacks tested basic sensations in her hands, such as her response to touch or temperature, her responses were normal. So, Dr. Sacks told Madeleine's nurses to place her food near her table but not to feed her immediately. One day, Madeleine, hungry and impatient, groped for her food, found a bagel, and brought it to her mouth. "This," Sacks writes, "was her first use of her hands, her first manual act, in sixty years." What followed was a great awakening as Madeleine began to explore the world with her hands. She found that she could use her hands not just to perceive but also to express her feelings. Soon, she asked for clay and began to sculpt, and within a year, she had become locally famous as the Blind Sculptress of St. Benedict's.

I was deeply moved by this story, an account of an individual who discovered a new way of sensing and exploring the world very late in life. I felt a tremendous connection with Madeleine. Both Madeleine and I did not develop normal sensory skills as babies and, as adults, had to be prompted, even provoked, into discovering how to use our hands or eyes. Once we did, we crossed a threshold. The Brock string was for me what the bagel was for Madeleine. As Madeleine handled the bagel, she learned how to perceive the world through touch, and as I learned to coordinate my eyes, I learned how to explore the world anew through vision. Describing Madeleine's story, Sacks asks, "Who would have dreamed that basic powers of perception, normally acquired in the first months of life, but failing to be acquired at this time, could be acquired in one's sixtieth year?"

I loved my new vision and wanted to tell everyone all about it, but after several embarrassing moments, I gave up trying. One day, I was having lunch with a good friend who was, as usual, going on about the politics of the day. I interrupted her passionate monologue with my own stream of consciousness:

"See that chair over there?"

She stopped abruptly in mid-sentence, somewhat annoyed, but allowed me to continue.

"Look at that chair," I said, pointing just outside the restaurant window. "The seat of the chair is wider than I used to think—and deeper too. There is actually more room to put your bottom on that chair than I ever realized." My friend looked at me with a mixture of confusion, amusement, and alarm, then went back to talking politics, which she must have thought was a safer subject.

A week later, I was walking across the Mount Holyoke campus on a quiet, weekend afternoon. There had been a snowstorm the night before, and the campus was blanketed with a new layer of clean, bright snow. In one area, large branches had fallen off a tree and lay in a high, tangled mass over the snow-covered lawn. Here was a place I could get my 3D fix for the day. I looked around; no one was present so I lay down on my back and slid myself under the jumble of branches. Looking up, I could see a lovely three-dimensional network, and I spent several minutes savoring the view. As I wiggled myself out of the branch pile, stood up, and dusted off the snow, I noticed one of my colleagues standing a little ways away, staring at me. He had a smirk on his face.

"Sue, what the heck were you doing?"

I didn't know how to describe succinctly to him the change in my vision, so I told him instead that I was just looking at the way the branches were growing out of the ground. I felt like an idiot.

In a patronizing tone, he pointed out that the branches weren't growing out of the ground but had been blown down in the storm. "You're a biologist." He added, "You should know that."

With an incident like this, I wondered whether I was going nuts. Perhaps I was simply delusional, making up a new way of seeing because I wanted to view the world like others did. Clinical tests in Dr. Ruggiero's office indicated that I was indeed gaining stereovision, but, with all I had read about the critical period, I was still unsure. I kept looking at the world through one or both eyes, constantly testing my new vision. I was completely distracted at work. In lectures, I actually paid far more attention to the empty, but now palpable, space I could see between the rows of students than I paid to the content of my talk. One day in neurobiology class, however, my lecture theme and my preoccupation came together.

When I learned to see in stereoscopic depth, I realized that wiring must have changed in the visual cortex of my brain. This is the area of the brain where neurons are found that give us our sense of stereopsis. Changes in brain circuitry depend in part upon the growth of new connections, or synapses, between neurons and the elimination of old ones. But the fastest learning may result from changes in the strength of already existing connections. My first 3D view, seeing the steering wheel pop out in front of the dashboard, happened very quickly after a long vision therapy session with the Brock string. This dramatic effect came on so quickly that it may not have resulted from the growth of new connections. Instead, it probably arose from a change in the relative strength of connections from my two eyes onto my visual cortical neurons, a change brought on by practicing my vision therapy procedures.

In learning, as with life in general, timing and balance are everything. How strong or weak an individual synaptic connection becomes depends upon when and with whom it is active. I was describing this idea to my class when I realized that this process might explain the changes I was now experiencing. I drew on the blackboard a circuit involving three nerve cells, a circuit that might exist in the visual cortex of my brain. One neuron belonged to the right-eye visual pathway, and a second belonged to the left-eye pathway. Both of these neurons synapsed onto a third cell that I called the postsynaptic cell. The right-eye neuron made a strong connection with the postsynaptic cell so that every time the right-eye neuron fired a nerve impulse, it stimulated the postsynaptic neuron to fire as well. On the other hand, the neuron from the left-eye pathway made only a weak connection with the postsynaptic neuron. When the left-eye neuron fired, it produced a small excitatory response in the postsynaptic cell, but not one strong enough to stimulate the postsynaptic cell to fire an impulse. If a scientist had been recording from the postsynaptic cell, he would have noted that only the right-eye input stimulated this cell to fire and would have classified this neuron as monocular.

However, there is a way to strengthen the weak connection between the left eye and the postsynaptic cell. If the left-eye neuron can fire at the same time as the right, then it will stimulate the postsynaptic neuron simultaneously with the right-eye cell. Under these conditions, the right, left, and postsynaptic neuron all fire at the same time. By a process called long-term potentiation,

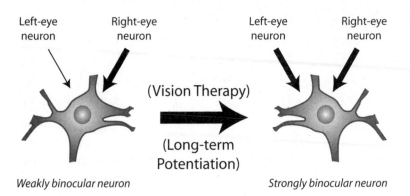

FIGURE 6.4: The postsynaptic cell is the binocular neuron. The thickness of the arrows indicates the strength of the connections. (© Margaret C. Nelson)

the previously ineffective connection between the left-eye pathway and the postsynaptic neuron can be strengthened, perhaps to the point where stimulation of the left-eye neuron alone can get the postsynaptic cell to fire. The postsynaptic neuron will then no longer be labeled as monocular but as binocular instead (Figure 6.4). It all comes down to the timing.

Now I had a possible explanation for my own visual changes. When I was an infant, neurons in my visual cortex probably received synaptic connections from both the right- and left-eye pathways. Since my eyes were misaligned, however, the input from the two eyes was decorrelated. A given visual cortical neuron received different information from the two pathways or perhaps received similar information from them but at different times. As a result of this asynchrony, the input from one of the eyes probably dominated, while that from the other eye grew weaker. Although the input from one eye to a given neuron may have become very weak, it may not have been lost entirely.

So, when I began vision therapy, many of my visual cortical neurons may have received strong input from one eye but only very weak and ineffective input from the other. Though my childhood surgeries brought my eyes into closer alignment, I still didn't know how to coordinate my two eyes for stereovision. As a result, similar inputs from the two eyes rarely impinged on a cortical neuron at the same time.

With the Brock string, I learned something that most six-month-old infants have already mastered. *I learned to aim both eyes at the same place at the same time.* Now a binocular neuron, even a very weakly binocular neuron, received correlated input from the two eyes. This situation could have altered the strength of the synapses on my visual cortical cells. Whereas input from one eye had always caused a given cortical neuron to fire, now it fired in response to the combined right- and left-eye inputs so that the pathways from both the right and left eyes were strengthened. If wired to detect retinal disparity, this neuron could now provide me with clues to stereoscopic depth.

I could think of additional mechanisms that might contribute to the change in my vision. For example, one eye may have blocked or inhibited connections from the other eye onto a visual neuron, and this inhibitory effect may have been reduced when I was able to aim both eyes accurately at the same place in space. As I gave my lecture and thought out loud about possible mechanisms for my visual transformation, I felt more confident that I was indeed developing stereovision.

———

When I first learned about stereopsis in college, I wondered if I could imagine this way of seeing. Now I had my answer. I could not. Stereopsis provides a distinctive, subjective sensation, a quale.

In their book, *Phantoms in the Brain,* V. S. Ramachandran and S. Blakeslee define the term *quale* (plural: *qualia*) as "the raw feel of sensations such as the subjective quality of 'pain' or 'red' or 'gnocchi with truffles.'" While I could infer indirectly a sense of depth through cues like perspective and shading, I could not synthesize stereoscopic depth from other visual attributes, such as color, position, form, or brightness. The sensation provided by stereopsis of empty space and things projecting or receding into that space is unique.

Just as I could not imagine a world in stereo depth, an individual with normal stereopsis cannot experience the worldview of a person who has always lacked stereopsis. This may be surprising because you can eliminate cues from stereopsis simply by closing one eye. What's more, many people do not notice a great difference when viewing the world with one eye or two. When a normal binocular viewer closes one eye, however, he or she still uses a lifetime of past visual experiences to re-create the missing stereo information.

I became convinced of these ideas on the day that the Star Wars movie *Revenge of the Sith* opened in movie theaters. My husband and children insisted that we attend the first screening at midnight. I like Star Wars, but until that night, I couldn't understand my family's fascination with the special effects. I was not thrilled with watching the new movie in the dead of night, but that evening, I saw something different. I was overwhelmed by the sense of space and volume created in the movie. Scenes of spaceships flying through the universe were fantastic! My new appreciation for the film didn't stem from my watching the movie in the wee hours of the morning or from significant improvements in cinematography since the previous Star Wars films. Instead, I was seeing the movie in a whole new way. Skilled cinematographers had used monocular depth and motion cues to create scenes on the flat, two-dimensional movie screen that suggested dramatic depth. Before my vision transformed, I could not experience this sense of space and volume while watching a movie because I had never experienced this sense of space and volume in real life.

Over and over again, I discovered that my theoretical knowledge of stereopsis did not prepare me in the least for the remarkable experience of seeing in stereo depth. I wanted to know whether others had written about this phenomenon. Although I read many books and articles by vision scientists and clinicians, I found that only the writings of Frederick Brock captured my thoughts and experiences. In fact, when I came across the passage quoted at the beginning of this chapter, I nearly fell out of my chair: "It must be repeated here that, before stereopsis is actually experienced by the patient, there is nothing one can do or say which will adequately explain to him the actual sensation experienced."

It is not possible for most scientists and doctors to experience both a normal and strabismic view of the world. Ophthalmologists, for example, must have excellent vision to operate on their patients' eyes. Yet, Brock seemed to have an uncanny understanding of how I once saw and how my vision had changed. Curious, I delved deeper into his writings and found out why. Frederick Brock was a strabismic. The first patient he had treated was himself.

7

WHEN TWO EYES SEE AS ONE

I want the unobtainable. Other artists paint a bridge, a house, a boat, and that's the end. They are finished. I want to paint the air which surrounds the bridge, the house, the boat, the beauty of the air in which these objects are located, and that is nothing short of impossible.

—Claude Monet

Two days after my first stereo experience, I was pulling clothes out of the dryer and noticed something quite odd. I happened to glance up at the collection of winter coats suspended from pegs fixed to a wall located about ten feet away from me. One of the coat sleeves was popping out toward me, capturing and defining a layer of space that I had never seen before. The folds and creases in the coats stood out in striking clarity and detail. The scene reminded me of a painting by Jean Auguste Dominique Ingres of the Princess Albert de Broglie (Figure 7.1). I had always liked the portrait but used to think that the pleats and grooves in the princess's clothes were painted in an exaggerated manner. No one, I thought, really saw the texture of the gown in that kind of depth and detail.

While our family's coats lacked the grandeur and eloquence of the royal gown, their wrinkles and folds were now as captivating

FIGURE 7.1: Princess Albert de Broglie, née Joséphine-Eléonore-Marie-Pauline de Galard de Brassac de Béarn, by Jean Auguste Dominique Ingres (1780–1867). (Image copyright © The Metropolitan Museum of Art/Art Resource, New York)

to me as the painting of the princess and her beautiful dress. I walked up to the coats and rearranged them, punching down the quilted material in some places, puffing it up in others, then walked away to see how they looked. I walked back and ran my hands along the material to see if the sense of three dimensions that I saw from ten feet away matched the three-dimensional feel and shape of the garments. I was reminded of my children when they were little babies, of how they found the wrapping paper around a gift to be more appealing than the gift itself. I also realized that I was calibrating my newfound stereopsis with my former ways of inferring depth.

Three-dimensional objects cast their images on our retinas, which are made up of flat sheets of cells. We construct a 3D view of the world from these 2D images using a combination of depth cues. With stereopsis, we compare the position of the images on the two retinas to construct a vivid sense of three dimensions. We also use cues that can be seen with just one eye. These are the same cues that painters like Ingres use to evoke a realistic sense of three dimensions on a flat canvas. In fact, by discovering and exploiting these cues, the artists of the past were in many ways vision scientists.

Imagine that you are a new, perhaps reluctant, student in a beginning art class, and your teacher hands you a lump of clay. She asks you to sculpt a ping-pong ball. You rub the clay between your two palms until its surface is smooth and round. Then, you paint the clay ball white. This proves to be no problem at all. You have created a three-dimensional sculpture to represent a three-dimensional object. So far, art class is a piece of cake.

Now, the teacher gives you a new assignment. You must draw the ping-pong ball. Again, you think that this will be ridiculously easy. You confidently draw a circle on a piece of paper.

But the picture you've drawn looks more like a flat, white disk. It could represent a cylinder seen from the top end or, perhaps, the bottom of a white, semicircular mound. Since you're trying to represent a three-dimensional object on a two-dimensional piece of paper, your picture is ambiguous.

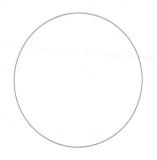

FIGURE 7.2

As you puzzle over how to make your circle look spherical (Figure 7.2), the art teacher suggests that you place your ping-pong ball below a small lamp. When light strikes the ping-pong ball from above, the upper surface appears bright, while the lower surface is in shadow. Since we normally see the world with light from above (from sunlight or from overhead lights in a room), we interpret a circular object that is bright on top and dark below as a sphere. So, you try making the top surface brighter than the bottom to represent this shading difference, and notice that your flat circle looks more 3D (Figure 7.3). This is one of the tricks Ingres used in his portrait of the princess. With shading, her gown looks so solid and real that you feel you can almost reach out and touch it.

If your lamp light hits the ping-pong ball from above but at a slight angle, the ball will cast a shadow—which it wouldn't do if it

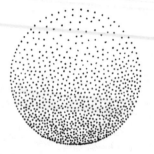

FIGURE 7.3. (© Julia Wagner)

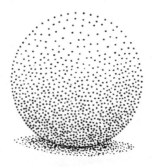

FIGURE 7.4. (© Julia Wagner)

were a disk lying flat on the ground. So, you redraw the ping-pong ball with a shadow as if it were illuminated from above at an angle, and now your drawing looks convincingly like a sphere (Figure 7.4).

The art teacher then shows you many additional tricks for suggesting depth, the same depth-perception cues that you unconsciously use in real life. If, for example, a tree in real life or in a painting breaks up a view of a fence, then you infer that the tree is standing in front of the fence. This cue is referred to as object occlusion. An artist can use this cue not only to suggest depth but to create an optical illusion, as M. C. Escher often did. In the Escher-like drawing in Figure 7.5, the artist has used the principle of object occlusion to confuse us. Both wheels are blocking parts of each other, so you cannot figure out their position in relation to each other.

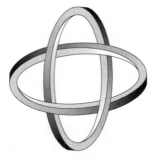

FIGURE 7.5: Which wheel is blocking which? (© Margaret C. Nelson)

FIGURE 7.6: Atmospheric perspective creates a sense of distance.

(© Malcolm Feinstein)

Due to the scattering of light as a result of the atmosphere, objects seen at a distance appear hazier than those in the foreground. Artists call this cue atmospheric perspective. Malcolm Feinstein, a professional artist who also happens to be my father, created the painting in Figure 7.6. Note how the more distant buildings and spire are painted to appear hazier.

Artists use rules of perspective to evoke a sense of depth. All of us learn through real-life experience that parallel lines, such as those outlining the sides of a long road, appear to converge in the distance. This cue is used often in paintings and illustrations. Take a look at how the use of perspective convincingly suggests depth and distance in the painting in Figure 7.7.

Since artists can evoke a strong sense of three dimensions on a flat canvas using monocular cues alone, I assumed that these cues were sufficient to provide me with a vivid, three-dimensional

FIGURE 7.7: A road painted in perspective.

(© Malcolm Feinstein)

view of the world. I used shadows and shadings to discern an object's shape, object occlusion to know which object was in front of or behind another, and perspective to gain a sense of depth and distance. I didn't need stereopsis to know that the world was in 3D. Gaining stereovision, I thought, would augment my perception of depth but not change it in any fundamental way. So, I was completely unprepared for my new appreciation of space and for the deep feelings of joy and wonder, the enormous emotional high, that these novel sights gave me. To experience for the first time seeing the most ordinary things in stereo feels like scaling a mountain and witnessing your first mountaintop view. Was I being overly dramatic, completely over-the-top? Had other individuals experienced similar visual transformations and responded with similar feelings? I wanted to discuss my story with other neuroscientists but was afraid that most would either discount my experiences as scientifically impossible or dismiss my reactions to gaining stereovision as exaggerated.

I thought back to Oliver Sacks's story "Hands" in *The Man Who Mistook His Wife for a Hat,* a true story about a woman who in her sixties had gained the use of her hands and a new way of perceiving and interacting with the world. I had met Sacks once, years before at a party, and we had talked briefly about my crossed eyes and my perception of the world without stereovision. So, one evening in December 2004, when my husband and children were involved in a late-night Monopoly game, I retired to a quiet room in our house and composed a long letter to Oliver Sacks. To my relief, Dr. Sacks did not dismiss my account but instead wanted to learn more about my vision change. He came to visit and, a year later, wrote an article, "Stereo Sue," for *The New Yorker* magazine. A week after the article's publication, I was interviewed on National Public Radio. The response to my story—hundreds of e-mails and letters—was incredible. I heard from individuals who, like me, had gained stereovision as adults, and all were more than anxious to tell their stories. I quickly learned that I was not alone, that they too were amazed, overjoyed, and sometimes baffled by their new vision.

When I gained stereopsis, borders and edges around objects appeared much sharper and crisper than ever before. This effect was almost as dramatic as my new sense of space. It is part of the reason I was so intrigued by the view of my family's coats hanging on the pegs outside my laundry room. An engineer would describe the world before my vision therapy as "low pass filtered," meaning that sharp edges were softened. In my stereoblind years, borders outlining objects were not well defined, but with no reference for comparison, I had no way of knowing this.

One of the people who wrote to me, Stephanie Willen Brown, put this very well when she described how objects, large and small, from letters on license plates, to items on her desk, and even to buildings in New York City, seemed sharp and clean after she gained stereovision. This clarity, she wrote, is "everything, everywhere. . . . There are edges to everything!"

Lucas Scully, who had been a patient of Dr. Ruggiero, experienced the same heightened sense of clarity, texture, and depth when, at age four, he received his first pair of bifocals to correct a visual condition called accommodative esotropia. Lucas was farsighted, which forced him to focus his eyes with great intensity to see close objects. When we look close, we not only focus our eyes but also turn them inwards. Since these two processes are coupled, some farsighted children overconverge their eyes when they focus on nearby objects. Bifocals reduce the amount of effort needed to focus and prevent the eyes from turning too far inward. When he wore his bifocals, Lucas's eyes didn't overconverge, and his stereopsis improved significantly. A letter his mother wrote to Dr. Ruggiero poignantly describes the effects of the new spectacles on her son's vision:

I will never forget the first day he wore glasses. Walking out to the car, after seeing his optometrist, Lucas was unusually distracted. I asked him several times to climb up into the truck but instead he squatted down and began running his little fingers along the pavement. My first reaction was to tell him not to touch. . . . My husband stopped me before I could say a word, and together we watched as Lucas experienced seeing texture for possibly the first time in his life. On the drive home he stared out the window watching the trees pass and at home his exploration continued. He walked around the house touching everything. He traced the grout between tiles with his fingers and spent half an hour looking at spices. He sat on the kitchen floor and pulled out jars of dried spices, held each one up to the light and examined the contents. . . . He went up to his room and took out toys and it was as if they were all brand new. He just kept touching them and turning them, inspecting each tiny detail. And that was just the beginning. . . . Now he walks up to children, introduces himself, and plays all sorts of games. As for coloring and painting, he can't get enough of it. . . . He loves puzzles. . . . He is writing. . . . Improving his vision has accelerated his social and cognitive skills. My shy, reclusive, difficult boy has become a curious, adventurous, and happy four-year-old.

Most of my friends and acquaintances who gained stereovision as adults experienced the same wonder that Lucas did when he got his glasses at age four. Rachel Hochman, who as an infant had a cataract of her right eye, expresses these feelings beautifully in an essay entitled "The Green World." A botanist who could always identify a vast variety of plants, Rachel felt that she truly understood the forest, even though she didn't initially have stereovision:

> Given names, these plants gained meaning. . . . These plants emerged from the green carpet and canopy. . . . The day I recognized that the world was no longer a sea of green, but rather a mass of specific individuals colored in olive, emerald, celadon, jade, teal, and chartreuse, brought me great joy. . . . The language of botany provided me with a context for perceptual distinction.

But Rachel was frustrated with her vision. She often squinted with her right eye, felt that there was a void in her vision on the right side, and had a hard time visually tracking objects like a moving ball. After hearing about my story, she embarked on a course of optometric vision therapy with Dr. Hans Lessmann. For the first weeks, Rachel worked tirelessly on tasks that taught her to pay attention to the input from her compromised right eye while her left eye was still open. "Every procedure," Rachel wrote to me, "required my intense, focused, concentration. I couldn't work while other people were in the clinic. I couldn't hold a conversation [while doing the exercises]. I can't believe how extraordinarily difficult it was for me." But the effort paid off. After six weeks, Rachel was ready to learn to fuse images from her two eyes.

To teach fusion and the use of stereopsis cues to Rachel, Dr. Lessmann had projected a Polaroid vectogram onto a wall so that images on the vectogram covered most of Rachel's visual field. A vectogram consists of two clear Polaroid sheets, each containing a similar image. Figure 7.8 shows the two sheets partially overlapping each other. Each sheet contains an image of a rope circle (also known as a quoit). When the viewer is wearing Polaroid glasses, each eye sees the image on only one of the sheets. To see just one image of the rope circle, he or she must fuse the right- and left-eye views.

When the images of the rope circles were projected onto the wall, they provided such a large target that Rachel was able to fuse the two images. Fusing large images, called peripheral fusion, is an important step in gaining stereopsis because it helps align the eyes and trigger accurate convergence and divergence movements. When Rachel could fuse the two images into one, Dr. Lessmann slid the vectogram sheet seen by the right eye to the left and the vectogram sheet seen by the left eye to the right. Now Rachel's eyes had to turn in to maintain fusion, to continue seeing just one combined image of the circles. To Rachel's astonishment, this image of the fused quoits appeared to float forward. If Dr. Lessmann slid the sheets in the opposite direction, then Rachel's eyes had to diverge to maintain fusion, and the fused quoits appeared to recede. She saw the fused image as being located no longer on the wall but in its own plane with a space, a palpable volume of space, between itself and the wall.

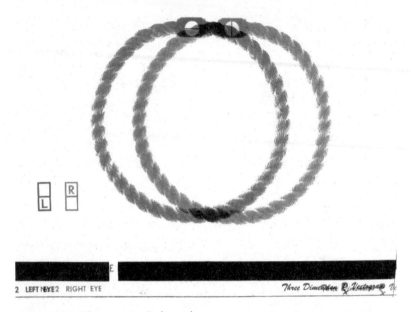

FIGURE 7.8: The quoits Polaroid vectogram. (Photo by James Gehrt)

The very next day, Rachel experienced her first stereo view outside of the doctor's office. She observed a familiar row of trees that were now completely transformed:

> I thought that I could see plants clearly. . . . I thought I knew the forest . . . [but] strolling through a humble plantation of pines, I had an epiphany. I was amongst the trees, not looking out at them. . . . [T]hey surrounded me in a way that was different than I'd previously experienced. The crenulations of bark and appliqué of moss were deeper, the edges clearer, the colors brighter. . . . Most unusual: the space between the trees was apparent. It was as if I had stepped inside a painting that I had spent my whole life observing. I was awed and moved to tears. I had never experienced a forest in this way. The depth of space and emotion was overwhelming.

Intrigued by Rachel's description of the large, floating rope circle projected into space, I traveled to Dr. Lessmann's office in Pittsburgh to try out this technique. Dr. Lessmann projected the rope circle onto the wall while I wore the Polaroid glasses and fused the right- and left-eye images into one.

"Where is the rope circle?" he asked.

"I'm not sure," I said hesitantly. "I think it's on the wall."

Dr. Lessmann handed me a long pole and instructed me to place the tip of the pole in the center of the rope circle so that it touched the wall. I put the tip of the pole in the center of the rope circle, but I couldn't feel the wall. It was disorienting—an uncanny feeling, much like the sensation you have when you descend a staircase and underestimate the height of a step. The wall seemed to have slipped behind the rope circle. Gingerly, I moved the pole through the image of the rope circle into empty space. Eventually, I hit the wall.

"Give the wall a hard tap," Dr. Lessmann commanded.

I did, and, in a single moment, everything changed. The rope circle shrank in diameter and appeared to float in front of the wall. I had made a perceptual shift. I knew intellectually that the rope circle was projected onto the wall and that the wall was a stable, unmoving object. But in order to fuse the right- and left-eye images, my eyes were aimed not at the wall but at a location in front of the wall. When I tapped the wall and used my sense of touch to gauge its distance from me, I learned that it was not at the distance where my eyes were directed, not at the depth of the fused image of the rope circle. I had to come up with a new interpretation—the rope circle was floating in space in front of the wall.

A normal binocular viewer makes this sort of interpretation all the time. But this way of seeing is new to someone still learning how to use stereopsis cues to see in depth. As Dr. Brock noted in several

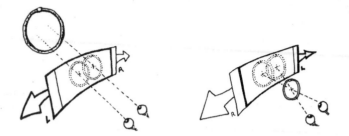

FIGURE 7.9: As the fused image of the rope circle appears to recede in space, it looks larger. As it appears to float forward, it looks smaller. (© Julia Wagner)

of his articles, "A strabismic speaks a different language." With procedures like this, I learned to speak the language of stereovision.

I experienced another curious effect with the rope circle vectogram that optometrists call the small in, large out (SILO) phenomenon (Figure 7.9). When I first saw the rope circle float forward in Dr. Lessmann's office, the rope circle also appeared to shrink in size. When the rope circle appeared to recede away from me, it also looked larger. This was counterintuitive as I knew that something coming closer to me should appear larger, while something moving away should get smaller.

A phenomenon called size constancy helps to explain this sensation. Size constancy allows us to judge accurately the size of people and objects located at different distances. If you look at a person across the room from you, he doesn't look tiny; he still appears life-size. Yet, if you use your fingers to outline your view of the person in order to determine the amount of space that he takes up in your visual field, you will notice that the amount of space is really quite small. Somehow, you have adjusted for the fact that the image of the individual will take up less and less retinal space as the individual moves away from you.

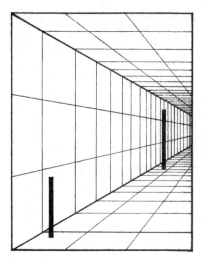

FIGURE 7.10: The corridor illusion. (Deregowski J in Gregory RL, Gombrich EH (ed), 1973. *Art and Illusion,* London: Duckworth)

The phenomenon of size constancy altered the way I saw the fused image of the rope circle in the Polaroid vectogram. As Dr. Lessmann slid the two sheets toward each other, it appeared, thanks to stereopsis, that the rope circle was coming toward me. But I was fooled. The rope circle was still projected onto the wall a fixed distance away. The image of the rope circle on my retinas had not changed size. It was only my newfound ability to use retinal disparity cues that gave me the illusion that the rope circle was closer. If the fused image had been a real object actually moving toward me, I would have unconsciously made an adjustment for the size of that image so that the object wouldn't appear to enlarge as it approached. So, when the rope circle appeared to move closer to me, I saw it as smaller as well.

The "corridor illusion" also makes use of the phenomenon of size constancy. The two poles in Figure 7.10 are actually the same height. However, we judge the pole on the right to be further away due to the decreasing size and converging lines of the floor, wall, and ceiling tiles. Since we interpret the right pole as further away, we see it as taller. Even if we know the cylinders are the same height, it's incredibly difficult to view them as such.

The SILO phenomenon helps optometrists to know whether a patient is seeing with stereopsis. The doctor will slide the sheets of the vectogram apart and ask the individual to report any size changes that occur, along with any sensation that the image is floating toward or away from them. If the patient reports the appropriate size changes, then the optometrist knows that he or she is using retinal disparity cues to see.

———

When I got back to Massachusetts after my meeting with Dr. Lessmann, I purchased a set of vectograms for my own use and worked at seeing the rope circle float in space. Initially, this was quite difficult. Most people learn to integrate these cues as infants. They begin to see in stereo at a time when they first start swiping at those brightly colored toys hanging from their cribs and strollers. With their first reaching movements and initial attempts to crawl, they learn to combine monocular and binocular cues to depth and then hone these skills throughout the first ten years of life. For me to see the "float," I had to learn to trust my new stereopsis cues and combine them with my old monocular ways of inferring depth.

As I made increasing use of my stereopsis cues, I was often surprised by new sights. In the past, any view framed by a window or mirror always appeared at the plane of the frame. Now, when looking out a window, I can see the space between the window pane and a tree outside. At first I was startled when looking in the mirror to see my own reflection not at the plane of the mirror but some distance behind it. I still get a kick out of walking back and forth in front of the mirror, watching my own reflection recede and come forward in the reflected space.

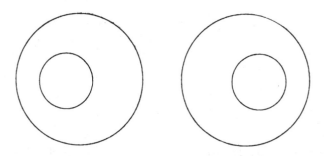

FIGURE 7.11: A non-random dot stereogram. If you are able to cross your eyes and "free-fuse" these two images, the inner circle will appear to pop out. If you fuse the images by looking "though" the page, the inner circle will appear to recede behind the paper.

Although I saw the steering wheel of my car "pop out" in 3D in 2002, I did not see depth or hidden images in random dot stereograms until much later. I was quite happy about "getting" some random dot stereograms because many scientists believe that seeing images in these stereograms, images that cannot be seen with monocular cues alone, is the ultimate proof that a person has stereopsis. In the non-random dot stereogram shown in Figure 7.11, it is easy to see how the fused image will consist of two circles. However, in a random dot stereogram, the stereo pair consists of an array of confusing dots that do not form an image unless they are seen by both eyes. If you cross your eyes to free-fuse the stereo pair in Figure 7.12, a central square will appear, as if by magic, to float in front of the background. (Free-fusing a random dot stereogram is difficult to do even for people with normal vision.)

Once I was able to "get" some of the easiest random dot stereograms, I found that I could also see some of the hidden images in the popular Magic Eye books. To see the hidden images, I had to override monocular cues that told me I was looking at a flat

FIGURE 7.12: A random dot stereogram. (Stereogram by Benjamin Backus)

drawing and place ultimate trust instead in what my stereopsis was telling me.

––––––––

Learning to see with stereopsis changed more than my sense of empty space. It altered my perception of being in space and moving through it. Since I had had imperfect vision, I relied more on the way my body felt and less on what I saw in order to get around. This became clear to me several years ago when on vacation with my family at a tropical resort. Every day, we would walk along a landscaped and terraced path from our cabins to the dining hall. One night, the electricity went out, and we had to make our way back to the cabins by the dim light of a crescent moon. I was surprised to discover that everyone but me was hesitant about finding their way back. Even though I was the one who usually got lost, I confidently led my family back to the cabins. I realized that without knowing it, I had memorized the route, not by counting my steps but by learning the timing and rhythm of my body's movements as I made my way from one location to the next. When we came to an outside staircase, I knew how many steps we needed to descend because of the familiar rhythm of my own steps as I went down the stairs. Back home from vacation, I understood why my children always turned on the light when they climbed up the stairs.

What was true for me was also true for my stereoblind friends. Many have told me that they are dedicated dancers and ice skaters, but very few described a passion for playing ball. Indeed, individuals with binocular vision disorders tend to enjoy activities in which they feel their body move through space rather than watch and interact with things in motion. Not surprisingly, my favorite activities in the past had been swimming and snorkeling, activities in which I felt myself enveloped in the water and had a strong sense of my body in motion.

When I gained stereopsis, I felt like I was immersed in a medium more substantial than air, a medium on which tree branches, flower blossoms, and pine needles floated. I wondered if this sense of the air was what Monet spoke about in the quote at the beginning of this chapter: "I want the unobtainable. . . . I want to paint the air." Or perhaps Eric Woznysmith, a strabismic, echoed Monet's thoughts when he described what it was like for him to see with stereopsis. Eric had studied drawing and learned that artists pay attention not just to the objects they will draw but also to "negative space," that is, the space, or the air, to the sides, in front of, and behind objects. When he gained stereovision, he told me that he could see one hundred times more negative space. Indeed, when I walk in a forest these days, I pay more attention to the pockets of space between the branches and trees than to the trees themselves. I seek out particularly beautiful volumes of space and like to immerse myself in their pockets.

This new sense of immersion in space is completely captivating and enchanting. Indeed, Rachel Cooper, who had a form of amblyopia (lazy eye), notes that before gaining stereopsis, "It felt like I was here and everything I was looking at was over there. I couldn't visually perceive or measure the space between me and other objects." Now, Rachel sees in 3D, and she says, "It feels like I am in the world. Empty space looks and feels palpable, tangible— alive!"

Rachel's words remind me of a wonderful sight I experienced one late winter day in the early stages of gaining stereopsis. I rushed out of the classroom building to grab a quick lunch, and I was startled by my view of falling snow. The large wet flakes were floating about me in a graceful, three-dimensional dance. In the past, snowflakes appeared to fall in one plane slightly in front of me. Now I felt myself in the midst of the snowfall, among all the snowflakes. Overcome with happiness, I forgot all about lunch and stood quite still, completely mesmerized by the enveloping snow.

————

Initially, I couldn't understand how my acquisition of stereopsis could have evoked this powerful sense of immersion in my surroundings. According to textbook descriptions, stereopsis provides an increase in depth perception only for objects at the distance at which your two eyes are aiming. But my whole sense of space had changed. Heather Fitzpatrick, strabismic since age two, began vision therapy with optometrist Carl Gruning. She echoed my feelings when she described her first experience seeing in 3D:

> The coolest thing is the feeling you get being "in the dimension." It is alive and open and you can actually see things floating by you as you walk and the depth is everywhere. . . . [I]t is ahead, but it is also down towards my feet. . . . [T]ables looked really low and walls looked really high and sitting at a desk I just wanted to put my hands all over it and push my hand in between the spaces of the objects on the desk.

To understand my new sensations, I took a careful look at some of the older scientific studies of binocular vision and stereopsis. Here I found an explanation for why I was seeing my surroundings in layers and layers of depth. As I describe in chapter 6, objects located in front of and behind the plane at which you are looking cast their images on your retinas outside Panum's fusional area. Thus, the images fall on regions of your two retinas too disparate to be fused. If you have normal vision, however, you unconsciously make a judgment as to whether or not the images come from the same or different objects. If the images come from the same object, then you unconsciously compare where the images fall on the two retinas and use this information to interpret how far away this object is. Although the estimate of the object's exact location or depth is not precise, you have an impression of its "nearerness" or "furtherness." It was this newfound sense of stereo depth for objects all around me that gave me the powerful feeling of being enveloped by the world.

But there were even more surprises in store: I started to see far more depth while moving, a phenomenon called motion parallax. As you sway, for example, to the right, objects near you appear to move to the left, while more distant objects appear to move with you to the right, and the closer objects seem to move at a faster rate than the distant ones. This difference in the relative motion of objects provides you with one of the most important depth-perception cues.

Since motion parallax can be seen with one eye, I, like most other scientists, assumed that I had always obtained a good sense of depth through motion parallax. Then, one day in the fall of 2005, I learned that I was mistaken. I had based this assumption on the erroneous premise that a strabismic sees the world like a person with normal vision who has simply closed one eye.

On that fall day, I was taking my dog for a walk along our usual route. My schnauzer felt compelled to sniff every blade of grass, leaving me bored and impatient, swaying absentmindedly underneath a dense network of tree branches. While gazing up at the trees, I was startled to see the branches in layers of depth. I saw how the outer branches captured and enclosed a palpable volume of space into which the inner branches permeated. I could make sense of the whole intricate network. From that day to the present, I have always walked to work so I can pass under the trees and get this stunning sense of three dimensions.

Since I was surprised by my improved sense of depth through motion parallax, I went to the library to read more about the topic. I learned that the same neurons and circuits that give us stereopsis may also provide us with our sensation of depth through motion parallax. So, my ability to see in stereo also translated to a heightened sense of depth through motion.

What's more, experiments by neurobiologist Mark Nawrot and his colleagues at North Dakota State University have identified the signals coming into the brain that provide us with our sense of depth through motion parallax. These studies reveal that pursuit movements of the eyes allow us to judge depth in this way. To keep my gaze fixed on one spot as I swayed under a tree, I unconsciously moved my eyes right when my head moved left and vice versa. My brain compared the direction of these eye movements with the way the different branches appeared to move. If a branch looked like it was moving in the same direction as my eye movements (that is, opposite to my head movement), then it was perceived as being closer to me and vice versa.

The same researchers have also reported that individuals with crossed eyes and amblyopia have a poor sense of depth through motion parallax. These results make sense to me because crossed eyes in infancy can impair pursuit movements. As my pursuit movements improved with vision therapy, I could make better use of motion parallax.

However, I don't think these studies provide the whole explanation. The impression I had of the volume or space between the branches was a recent sensation for me, one provided by my new-found experience with stereopsis. Combine motion parallax with the sense of palpable space from stereopsis, and the effect is dramatic. For me, as well as for many of my formerly stereoblind friends, one of the greatest surprises and delights of our new vision has been this incredible sense of depth while moving.

But the visual surprises did not stop there. A few months after my walk with the dog, my daughter and I decided to watch the film *Legally Blonde* on her laptop. During the opening credits, the movie camera swept across a large vase of flowers, and I immediately saw the whole collection of flowers in dramatic depth.

"Whoa," I said to Jenny, "did you see that?"

"See what?" Jenny asked, but I had already rewound the movie to see the scene again. I would have watched the scene several more times if I'd thought Jenny would tolerate it. When the cameraman swept the camera around the vase, I saw the flowers on the screen as if I were rotating around them and this gave me a great sense of 3D. This sensation is similar to the ability to see "structure from motion," that is, to envision the 3D form of a solid object by watching its moving shadow. So, I went back to the library to read about structure from motion and was not surprised to learn that the capacities to see depth through stereopsis and to determine structure from motion are linked. People with poor stereopsis have a poor sense of structure from motion. As I gained stereopsis, I could see more depth while watching videos. Even though movies and TV shows are displayed on flat screens, I actually enjoy watching them much more now than I did before developing stereovision.

———

As much as I loved my newfound ability to see the empty space around me, my new sense of immersion could be unnerving and downright frightening at times. On a trip to Hawaii with my husband and children in the early stages of my stereo life, I visited a scenic viewing spot overlooking a beautiful canyon. I went right up to the protective railing to take in the view. Suddenly, I felt like I was floating unsupported high above the incredibly deep canyon. I quickly backed away from the railing, moving to a safer spot behind other tourists and the view below. Later on that day while hiking, I felt panicked every time my children or husband approached the cliff's edge. I had to use all of my self-control not to scream at them to get away from the cliff.

When I returned home, I realized that my stereovision contributed to this new fear of heights. I was standing on a familiar bridge overlooking a waterfall on the Mount Holyoke campus when, again, I felt this new and unnerving sensation of floating unmoored, this time above the water. Since the bridge was not high and the scene very familiar to me, I could control my panic, relax, and eventually enjoy the sensation.

Despite my experience in Hawaii, my adjustment to a world in stereo went fairly smoothly because my worldview changed gradually. First, I saw more depth only for close objects and then gradually for objects further away. I had a chance to savor each new sight and time to learn how to move through an increasingly depth-filled world. For some people, however, stereopsis emerges suddenly and in full measure, and this abrupt change can be very disorienting and frightening.

When Cyndi Monter, a forty-eight-year-old woman with strabismus and amblyopia, was attempting to eat her lunchtime salad, her 3D vision "clicked in." Suddenly, she was looking at a salad in which tomatoes popped out at her and the lettuce leaves appeared to lie in very distinct layers of depth. Cyndi was so alarmed

that she couldn't eat her food and headed straight to her optometrist instead. As she drove, roadside trees and signs seemed
to loom threateningly toward her. Her optometrist, Carl Hillier,
ran some tests and informed her that she was now seeing in stereoscopic depth. He had tried to warn her that her work in vision
therapy might bring about these changes, but without any prior
experience with stereopsis, Cyndi couldn't imagine what he meant.

Both Jennifer Clark and Tracy Gray experienced an abrupt onset of stereovision as well—and with it a wide-angled view of the
world. Tracy wrote to me about the moment when her vision suddenly transformed:

> I was sitting on the sofa in my living room about midnight,
> and I immediately saw the whole room and all the objects in it
> in 3D. I was able to take in so much more of the room than I
> did before (previously, I would generally fixate on whatever ob-
> ject I was looking at; I didn't have tunnel vision . . . but I don't
> think I really took it all in). When I got up from the sofa and
> walked around, for the first time I was walking through the three-
> dimensional space, clearly walking among the objects. It was so
> overwhelming and a little disorienting, that I went straight to
> bed. When I woke up, I was wondering if it was still going to be
> there, and when I opened my eyes, I saw the *whole* room, and
> my ceiling fan was sticking out.

Jennifer Clark, who underwent vision therapy with optometrist
Caroline Hurst of St. Neot's in England, notes the following:

> I had about four months of eye exercises with initially only some
> small effect. Then suddenly my vision changed to 3D completely,
> in one shot, whilst walking down the corridor at work one day.
> It was very dramatic as my peripheral vision suddenly filled in

on both sides and the corridor ceiling changed completely from a parallelogram to a rhombus.

For weeks, Jennifer struggled to integrate her new views with her old ways of interpreting the world. She would run her hands along the TV set to reconcile its three-dimensional feel with its new three-dimensional look. For many months, she was plagued with motion sickness. Buildings appeared to lurch up and down as she walked along the road. With more therapy, Jennifer's world-view stabilized, and she began to relish her new way of seeing.

Similarly, Tracy initially felt disoriented and overloaded:

It took a long time to integrate it [stereovision] and have it not be overwhelming. [My view of the world] was too bright, and my eyes watered for several days. [My optometrist] recommended no television, no reading, no driving, and as little work at the office as possible, and to work to integrate it by walking outside in short segments and trying to take in the whole scene. For the first several days, I was only able to stay at work a few hours a day. I had to spend a fair amount of time each day just lying down with my eyes closed. It was about a month before I drove a car again. . . . With my vision like it is now though, I can shut one eye and still see more than I used to with two.

Tracy's comments helped me to understand my own experiences, for there were many days when I felt both exhilaration and exhaustion. My old way of seeing had given me the world in smaller doses. With my new outlook, all of my senses were awakened. The world was keenly present. While I used to play the piano once a week, I started to play every day, pausing often in the middle of a piece, captivated by a given theme or combination of musical intervals. Like my view of the coats mentioned at the beginning of this chapter, I saw much more texture and shape in the objects around me. I felt as if I could touch and manipulate everything with my vision alone and finally understood what philosopher Maurice Merleau-Ponty meant when he said, "Vision is the brain's way of touching."

But on many days, I also experienced sensory overload. I used to come home from work and immediately tune in to the radio to catch up on the news. In the early months of my vision therapy, I avoided the news altogether, listening to familiar music instead. I worried that I had lost concern for world affairs but later realized that I was perceptually tired. Now, my ordinary day was like a day in Disney World, a strange foreign city, or a tropical forest.

Once my vision changed, I was so preoccupied by my new sense of the world that I took on fewer responsibilities at work. I justified this by reminding myself that I was also caring for aging parents and growing children. Yet, in the past, I had willingly juggled many demands. What I was experiencing now was different. Most people learn to see when they are infants, at a time in their lives when they are cared for, are free to get cranky, and enjoy lots of naps. I was relearning how to see as a responsible, contributing adult. While I went through all the motions at work, I desperately wanted to be left alone, to be quiet and reverent, to take in one long, delicious look after another. I disappeared on long, solitary walks. I was at a loss as to how to explain this to my colleagues and friends.

———

Most surprising to me was that the change in my vision affected the way that I thought. I had always seen and reasoned in a step-by-step manner. I saw with one eye and then the next. When entering a crowded room, I would search for a friend by looking at one face, then another. I didn't know how to take in the whole room and its occupants in one glance. While lecturing in class, I always spoke about A causing B causing C. Until I watched my children grow up, I had assumed that seeing the details and understanding the big picture were separate processes. Only after I learned the details could I add them up together into the whole. I could not, as the saying goes, see the forest for the trees. But my kids seemed to be able to do both at the same time.

I remember a time when my family was given a tour of an old church's clock tower, and we came upon the mechanism that moved the clock hands and rang the church bells. Instantly, my son saw how the whole machine worked, the choreography of all of the gears. Similarly, when my daughter plays a sonata on her flute, she quickly grasps the whole piece, how the whole drama unfolds. The individual phrases and themes combine for her into one coherent flow. My son and daughter, when young, could grasp the details and the big picture at the same time. I didn't know how to do this until midlife, when I learned to see simultaneously with two eyes. Only then was I aware of the whole forest and, within it, the trees.

8

NATURE AND NURTURE

I dwell in Possibility—
A fairer House than Prose—
More numerous of Windows—
Superior—for Doors—

—"I dwell in Possibility" (poem 657), by Emily Dickinson

Seeing, as the saying goes, is believing, but during my first stereo-viewing year, I had a hard time believing what I saw. Although clinical tests in Dr. Ruggiero's office confirmed that I now saw in stereo, I was not totally convinced. Others also doubted my story. Some doctors argued that I must have had Duane's syndrome: most people with this disorder cannot move one or both eyes outward, although they can still see in stereo when looking straight ahead. But this was not my condition; I could always move both eyes freely in all directions.

Others claimed that my story was one in a million, yet I found other formerly stereoblind and strabismic individuals who had undergone the same remarkable visual changes. What set my story apart was that I was lucky enough to work with a developmental optometrist with expertise in binocular vision and optometric vision therapy. These eye doctors are relatively rare, so most patients do not know of or have access to such specially trained optometrists.

What's more, many adults with binocular vision disorders are told that their deficits are permanent, so they seek no further treatment. This is what I had been told before meeting Dr. Ruggiero, and it was what most of the strabismics who wrote to me had also heard. The sheer number of letters and e-mails that I regularly receive makes me realize how much my story has struck a nerve.

Arguments concerning the timing of treatments for crossed eyes bring up the age-old debate between nature and nurture. Is the brain at birth a blank slate, totally shaped by our experiences (nurture), or are we born with a brain prewired for certain mental abilities (nature)? In the late 1800s, two great scientists, Ewald Hering and Hermann Von Helmholtz, hotly debated the nature-versus-nurture question with regard to vision. These two men, who dominated the field of visual perception in their day, were as different in their temperaments as in their beliefs. Hering was forceful, volatile, and charismatic, demanding and receiving immense loyalty from his students. Helmholtz was cool, reserved, and modest. He invented the ophthalmoscope, a basic instrument used to examine the retina, and contributed to the fields not only of optics and perception but physics and philosophy as well.

Both scientists were keenly interested in the role of stereovision in spatial perception. Hering believed that our ability to locate objects in space was innate and an automatic consequence of where images fell on our two retinas. So, Hering saw "nature" as playing the dominant role. Helmholtz argued that spatial perception was not an automatic skill but was learned through experience. To perceive is a mental act, he wrote, an active interpretation of the input provided by our two eyes. Thus, "nurture," or experience and learning, plays a dominant role. The nature-versus-nurture debate between these two great thinkers is quite a difficult dilemma to resolve because a baby opens his eyes at the moment of birth. With a baby's first glance, visual experience influences the neuronal circuitry in his brain.

In the early and mid-1900s, nature, not nurture, was thought to play the major role in the development of strabismus. At that time, treatment for crossed eyes was heavily influenced by Claud Worth, an ophthalmologist and author of *Squint: Its Causes, Pathology, and Treatment*. Although we think of the word "squint" as a term for narrowing our eyes, it's technically another term for strabismus. Worth studied twenty-three hundred children with strabismus and found that most of these patients, even after surgery, never developed the ability to see in 3D. He concluded that children with infantile strabismus have a congenital weakness of the fusion faculty. In other words, surgery may appear to straighten the eyes of a strabismic child, but the child will never learn to converge or diverge his eyes properly because of his innate inability to fuse.

Francis Bernard Chavasse, one of Worth's students, went on to write new editions of *Squint*. But in 1939, thirty-six years after the first edition of *Squint* was published, Chavasse challenged Worth's theory by postulating that the ability to fuse depended upon a series of reflexes that developed during childhood. Surgery that realigned the eyes might allow for the development of these reflexes as long as the surgery was performed and the eye deviation corrected before the child was two years old. Chavasse opened the door for the role of experience, albeit only early experience, in the development of stereovision. Nature still dominated stereovision development, but nurture made an important contribution.

Scientific studies on neural development performed in the second half of the twentieth century later supported Chavasse's ideas. These experiments suggested that the newborn brain was not a blank slate. Instead, much of the circuitry in the brain was already established at birth or very early in life. For example, in a paper published in 1951, developmental biologist and Nobel laureate Roger Sperry demonstrated that neuronal circuitry between the eye and brain in fish and amphibians was "hardwired." In a mind-boggling experiment, Sperry removed the eye of a frog and reimplanted it upside down. After the operation, the frog would try to catch a fly buzzing above him by flicking its tongue not up toward the fly but down at the ground. Even after many failed attempts to catch the fly, the frog didn't change its behavior. Sperry's studies of frogs and newts suggested that connections between the eye and brain couldn't be altered in these animals and brought up the possibility that vision was hardwired in people as well.

Other behaviors in animals could be modified but only by experiences that occurred during a narrow time window shortly after birth. One such behavior, known as imprinting, was first noted by the naturalist Douglas Alexander Spalding in 1873 and described in detail in the 1930s by the great animal behaviorist Konrad Lorenz. Lorenz shared his house with some of his favorite subjects, the graylag geese. When geese hatch, they imprint upon, or accept as their mother, the first living thing that they see and hear. If Lorenz was the first creature that they saw and heard (he would quack and honk at his charges), then the geese "imprinted" upon him and followed him everywhere. A gosling who had imprinted on Lorenz never switched its allegiance to its real mother. Here was an example of a behavior that can be permanently changed but only by very early events. It was yet another clue that some functions are hardwired into the nervous system at or shortly after birth.

These were among the observations and ideas under discussion by scientists at the time that David Hubel and Torsten Wiesel began their studies of the mammalian visual system. Hubel and Wiesel did their first experiments together at Johns Hopkins University in 1958, moved their lab to Harvard in 1959, and continued to work together for another twenty-four years; theirs was one of the most fruitful collaborations in the history of the field

FIGURE 8.1: An example of a recording from a nerve cell taken from Hubel and Wiesel's first paper on the cat visual cortex. The lower bar indicates when the nerve cell was stimulated with a light pattern. During this time, the neuron fired impulses. (Hubel DH and Wiesel TN, 1959. Receptive Fields of Single Neurones in the Cat's Strate Cortex. *Journal of Physiology* 148: 574–91, Blackwell Publishing)

of biology. They developed techniques to record nerve impulses produced by individual neurons in an animal's brain, a procedure too invasive to apply to people. Recordings of nerve impulses resemble the waveforms seen on electrocardiograms and tell us when a visual neuron is active and responding to light (Figure 8.1).

Hubel and Wiesel studied neurons in the visual cortex of cats (as they were readily available) and monkeys (as their visual system is similar to our own). Most of their investigations involved neurons in the primary visual cortex, or the first cortical area to receive visual input. By observing the responses of individual neurons to patterns of light, Hubel and Wiesel uncovered some of the underlying neural mechanisms for vision. They examined, for example, how the activity of individual neurons allows us to discern borders and edges and appreciate different colors. Their work on very young animals also contributed to the nature-versus-nurture debate. Their studies supported the observations of Lorenz and other animal behaviorists who stated that the brain in early infancy is not a blank slate, a tabula rasa. To a large extent, the mammalian brain, including the visual system, is wired before or shortly after birth.

In the primary visual cortex, Hubel and Wiesel discovered binocular neurons. These cells respond with nerve impulses to input from corresponding areas of both retinas. A particular binocular neuron may receive input, for example, from the center of the fovea of each eye, while another neuron may receive input from an area, let's say, 2° to the right of and 3° below the fovea of each eye. The input from the two retinas may not be exactly the same, however. Since the two eyes are separated in space, they report a slightly different worldview. To see with stereopsis, you have to be sensitive to these small differences. Some of the binocular neurons in the cortex are sensitive to the small disparities in input coming from the right and left retinas. Hubel and Wiesel called these cells the binocular depth neurons, and they are important to our ability to see in 3D.

In the course of their remarkable twenty-five-year collaboration, Hubel and Wiesel went on to examine the effects of common childhood disorders such as strabismus and amblyopia on the development of visual circuitry. In 1965, they published a pivotal paper on strabismus describing experiments in which young kittens were made walleyed by cutting one eye's medial rectus muscle, which moves the eye inward toward the nose. As a result, the eye wandered outward preventing the young cats from fusing images provided by the two eyes. When the scientists made recordings from neurons in the visual cortex, they found hardly any binocular neurons at all. Instead, the neurons responded to only the right or left eye. They posited that the cats lacked stereopsis.

Many other vision scientists performed variations on Hubel and Wiesel's basic experimental paradigm, and these tests all produced similar results. Very young animals with artificially induced strabismus, it seemed, had a "monocular brain." Their neurons responded to one eye or the other but not both. Particularly striking to the scientists was that these effects were found only in young animals. If they made older cats strabismic, then their neurons remained binocular.

Hubel and Wiesel's research supported Chavasse's claim that surgery could promote stereovision in human strabismics only if it realigned the eyes by two years of age. Until the 1960s, surgery to straighten the eyes was not usually performed on strabismic children under the age of five. Partly as a result of the animal studies, surgeons began to operate on patients at progressively younger ages. They found that about half of the children developed some stereovision if their eyes were aligned and stayed aligned during the first year of life.

Hubel and Wiesel's research combined with the surgical results suggested that real-life experience, such as seeing with straight or crossed eyes, can alter brain wiring only during specific periods in early childhood. While we learn new skills throughout life, it was thought that basic perceptual abilities, like seeing in 3D, and their underlying neuronal circuitry were established in a permanent way in the first years of life. Indeed, it was my knowledge of Hubel and Wiesel's groundbreaking work on strabismus and visual development that caused me to repeatedly question my newly acquired stereovision. From my own study of synapses and learning, I could conceive of mechanisms that explained the change in my vision, but I didn't think I could convince others. So, it seemed best to go right to the source.

With considerable trepidation, I drafted a letter to Dr. David Hubel and received his response shortly thereafter. To my enormous relief, he didn't discount or dismiss my story. In fact, he felt that my stereovision would continue to improve as I continued to pursue vision therapy. In his letter and during our subsequent discussions, Hubel explained to me that he had never attempted to correct the strabismus in animals in order to examine the effects of straightening the eyes on visual circuitry. It would have been difficult to realign the eyes surgically and even harder to train the animals with vision therapy. So, he couldn't be sure that the effects of strabismus on binocular circuitry were permanent. Yet, to truly delineate a critical period, he and other scientists would have had to demonstrate that the effects of strabismus on cortical wiring cannot be reversed after a certain age. Indeed, David Hubel had already stated these concerns in *Brain and Visual Perception* when he wrote, "A missing aspect of this work is knowledge of the time course of the strabismus animals, cats or monkeys, and in the monkeys the possibilities of recovery."

Hubel suspected that binocular depth neurons are present in the visual cortex of newborn babies. Even though I had been cross-eyed since early infancy, I had not lost these neurons entirely. When I learned to move my eyes into position for stereovision, I finally gave these neurons the information they were originally wired to receive. So, nature (the innate binocular wiring of my brain) and nurture (my ability to relearn how to use my eyes) conspired to give me my new vision. Hubel's thoughts gave me confidence in my own reasoning, for I had considered the same ideas on the day that I described to my class possible mechanisms for how I may have learned to see in 3D.

Although we cannot record from visual neurons in newborn humans, laboratory experiments on animals suggest that binocular neurons are indeed present at birth. For example, binocular neurons sensitive to retinal disparity are found in one-week-old macaque monkeys, the earliest age at which recordings of nerve impulses can be made. Since strabismus is rarely, if ever, found in newborn humans, even infants with crossed eyes may be born with a normal complement of binocular depth neurons.

Although babies are probably born with binocular neurons, they do not demonstrate a capacity for stereovision until about four months of age, a concept that truly came to life for me last summer when I may have witnessed a young baby's earliest stereo moments. I was at a picnic, talking to a friend who was trying to entertain her infant and two-year-old child. The infant was happy in my friend's arms until she turned around to speak to me, at which point the baby began to wail. My friend looked exasperated and exhausted. "Let me hold the baby," I said. "I think I can make him happy." She handed me her son, and I took him, turned him around again, and let him glance at his leisure at the low-lying branches of a tree overhead. To my delight, the little boy stopped crying and stared wide-eyed at the branches. For the first time, I thought, this little boy may be seeing the space between each branch and how the branches project down toward him. "I know just what you're seeing," I said to the little guy. "It's spectacular, isn't it?"

If binocular depth neurons are present at birth, and if strabismus doesn't emerge until a few months of age, I wondered if cross-eyed infants initially have the capacity to see with stereopsis. So, I was intrigued to read studies published in the mid-1980s by Eileen Birch and David Stager of the Retina Foundation of the Southwest. These investigators performed the same "preferential-looking" experiments for stereovision on cross-eyed babies that they had previously performed on babies with normal vision. Since the visual fields for the two eyes of the strabismic infants do not overlap as normal, the babies in the experiments wore special prism goggles that aligned the visual fields of the two eyes. The investigators then presented the babies with two stereograms, one that looked flat and one that looked 3D. The babies could distinguish the flat from the 3D stereogram only if they had stereopsis. These cross-eyed children preferred to look at the stereogram in 3D. Even cross-eyed babies at four months of age can see with stereopsis. However, they lose this skill over the next several months.

In the experiments, the prism goggles made it possible for the cross-eyed babies to fuse images presented to the left and right eyes. Without the goggles, the visual fields of the two eyes were not well aligned, so the child experienced double vision and visual confusion. Under these conditions, the child will likely turn in one eye even more to make it easier to discount the input from that eye and obtain a single view of the world. As a result, binocular circuitry in the infant's brain will be weakened and masked.

If binocular circuitry does not disappear but is suppressed in a strabismic brain, there may be situations in which a strabismic can see in stereo. Indeed, several strabismic individuals have told me that they once saw briefly in stereoscopic depth. They never forgot these moments and recalled them in great detail. Richard Zack had been cross-eyed since early infancy, didn't have surgery until he was six years old, and was "stereoblind." But one day at work, when he was in his late forties, he wore shuttered stereo-viewing glasses to look at a 3D display that an optics laboratory was demonstrating. To his total amazement, he could see in stereo for the first time in his life, and the effect was immediate. Similarly, Steve Perez told me that he was also considered "permanently" stereoblind as a result of strabismus. He went to a 3D laser light show one night when he was in his teens and, with great effort, very briefly forced his eyes into position to see in stereo. His remarks jogged my own memory, because once as a child and once as a teenager, I was looking at a leaf at very close range and saw it popping out at me. I didn't know then that this was stereovision, but the views were so unexpected and astonishing that I still vividly remember those events.

These stories correlate with studies of human strabismics done by a small number of binocular vision scientists and optometrists. Their experiments demonstrate that strabismic individuals who flunk the standard stereopsis tests used in the eye doctor's office may be able to see in 3D when they look at large or moving targets located in their peripheral visual fields. Because of the strong belief in the critical period and the way animal studies have been done (see below), these observations have been overlooked by most other scientists and clinicians. Indeed, Frederick Brock noted, "Nearly all strabismics have occasional moments when they maintain binocular vision. The only reason this is not generally known is that most of us [eye doctors] have never taken the trouble to discover the fact."

One way that Dr. Brock and other optometrists promoted this latent stereovision in their patients was to project large stereo targets onto the wall. While the viewer is wearing Polaroid or red/green glasses, each eye sees only one target, and the individual must fuse the two members of the stereo pair into one virtual image. The stereo scenes that Dr. Brock used were so large that they covered a great deal of the visual field. Under these conditions, strabismics may have their first and best chance of fusing and seeing in stereo. With more time and practice, they can learn to fuse smaller and more detailed targets.

Vision therapy taps into the potential for stereovision that often lies dormant in a strabismic brain. Unfortunately, animal experiments on strabismus have been designed in a way that would never reveal this latent binocularity. The vast majority of these experiments have involved recordings of neurons that respond to only the central 5° of the visual field. These neurons have very small "receptive fields," meaning that they are activated by light falling on a very small area of the central visual field. In contrast, neurons that respond to less central stimuli are stimulated by light falling over a wider area. For a binocular neuron to combine input from the two eyes, its receptive fields for the left and right eyes must overlap. It is easy to see how eye misalignment will prevent overlap of the small receptive fields of central neurons but not of the larger receptive fields of more peripheral neurons. As a result, strabismus may affect binocular neurons in a central (small receptive fields) to peripheral (large receptive fields) progression. Experimental data, obtained almost exclusively from central neurons, may have exaggerated the effects of strabismus on the binocular nature of neurons throughout the visual cortex and underestimated the potential for stereovision in strabismic individuals.

In short, studies on human strabismics indicate that an individual who has some sight in both eyes is binocular to some extent. On one end of the spectrum are those who depend heavily on only one eye, while on the other end are people who use their eyes in close-to-perfect synergy. Optometrist Paul Harris calls this spectrum the binocular continuum. With vision therapy, people can increase their binocularity and shift their position along the binocular continuum. How far a person can move along this spectrum depends in large part upon how he has adapted to his visual disorder.

For some, vision therapy alone is not sufficient to achieve stable binocular vision. Pat Duffy had had a mild strabismus since early childhood, but her eye turn was intermittent and left untreated. Then one day when she was forty-two years old, she woke up with constant double vision. What's more, she couldn't look out of her strabismic eye alone without feeling nauseated. After brain scans and other tests had ruled out a critical problem, Pat began vision therapy with Dr. Ruggiero. A year later, however, she still saw double. There was one positive change, though—she could now comfortably look out of her strabismic eye alone. Dr. Ruggiero found an ophthalmologist for Pat, and she underwent surgery in the summer of 2007. When she woke up from the surgery, the double vision was gone. This was wonderful, but when she felt very tired, she saw double again. So, she went back to Dr. Ruggiero, and this time the vision therapy had a great effect. Pat now sees singly and in stereo.

Pat's story highlights an important issue in the treatment of strabismus. To solve her vision problems, she needed the expertise of two different types of eye doctors, an ophthalmologist who performed the surgery and a developmental optometrist who designed and administered the vision therapy. These two groups of eye doctors do not generally work together, so many people with strabismus do not get the kind of comprehensive treatment that they might need.

Even though I'd had corrective surgery, I continued to suppress the input from one eye. Surgery may have put my eyes in the ballpark for binocular fusion, but I needed the therapy to learn how to see in stereo. I started out toward the nonbinocular end of the binocular continuum and, with therapy, shifted my position toward perfect synergy. Nature and nurture, my innate circuitry and my capacity to learn, worked together to provide me with the vision that I now treasure.

———

The lazy eye myth:

Nowhere has the debate over nature and nurture and the concept of the critical period been applied more often than in the treatment of lazy eye, or amblyopia. This condition causes poor eyesight, usually in just one eye, that cannot be corrected by glasses or contact lenses and is not due to eye disease. Amblyopia affects approximately 2.5 percent of the population and causes more cases of monocular blindness in the under-age-forty population than all other eye diseases and injuries combined.

Caitlin Ronayne, now a high school student and patient of Dr. Ruggiero, has a form of amblyopia. Her left eye has turned in since infancy and had poor vision. In school, Caitlin had trouble switching her gaze from the book to the blackboard and was impatient with arts and crafts. When Caitlin was six, an ophthalmologist told her mother that her vision was permanently impaired, and nothing more could be done. Caitlin, it seemed, was destined for a life in which she saw the world largely through one eye.

Caitlin's ophthalmologist based his judgment partly on the results of scientific experiments pioneered by Hubel and Wiesel in the 1960s and 1970s. Hubel and Wiesel were aware that children born with a cataract of one eye did not recover functional vision in that eye after the cataract was removed. It seemed that early visual experience was important for normal vision. They wondered if this loss of vision resulted from a problem with the eye itself or with neuronal wiring in the brain. The best way to test this was to simulate congenital cataracts in an animal and then make recordings from visual neurons.

Hubel and Wiesel covered one eye of cats or monkeys shortly after birth. These animals were considered "monocularly deprived." After days, weeks, or even years, the covered eye was reopened, and recordings were made from the neurons in the visual cortex. In normal animals, most of the neurons respond to both eyes, but in the monocularly deprived cats, almost all the neurons responded to only the eye that had not been covered. If, while the animals were still very young, the situation was reversed so that the covered eye was opened and the open eye was occluded, then a change occurred in the cortical neurons. Most of them would switch their eye preference and respond to only the currently seeing eye.

As with the strabismus studies, the changes in eye preference occurred only during early infancy. If one eye is covered in an adult animal, no changes occur in the activity of the neurons. The same percentages of neurons still respond to both eyes. In fact, the changes in eye preference in monkeys are most dramatic if monocular deprivation occurs during the first three to six weeks of life. This stage was termed the "critical period," and the phrase was later used more loosely to describe the stage in early life when experience is thought to have its greatest effects on brain circuitry.

Hubel and Wiesel did not extrapolate their findings on monocularly deprived animals to the most common forms of amblyopia in children, but many physicians did. The majority of doctors and scientists assume that the visual system of most amblyopes has undergone the radical changes in circuitry observed in animals that have one eye covered shortly after birth. Since amblyopia rarely develops in people after the age of eight, they further surmise that visual circuitry cannot change after this age. They mistakenly assume that the "critical period" for the development of amblyopia is the same as the "critical period" for its rehabilitation. Brain plasticity, it seems, has an age limit. So, today, older children and adults with amblyopia are told that nothing more can be done. What's more, developmental optometrists who disagree with this conclusion and successfully improve vision in older amblyopes may be labeled as sharks and charlatans.

But do Hubel and Wiesel's monocular deprivation experiments really serve as a good model for the most common forms of amblyopia in people? These animal experiments mimic the effects of a cataract present in one eye of a child at birth, but this condition, called deprivation amblyopia, is very rare, affecting only 0.03 percent of children. Much more common are forms of amblyopia that develop from strabismus or a condition called anisometropia.

Amblyopia can be caused by strabismus when only one eye does the looking. If a young child looks through only one and the same eye at all times and turns in the other, then vision is reduced in the turned eye. This is what happened to Caitlin Ronayne and also to Cyndi Monter, the woman whose first stereo view was of her lunchtime salad. Strabismus is not present at birth but occurs at about two to three months of age, and amblyopia, which is an adaptation to the strabismus, develops sometime after that. So, unlike the animals in the monocular deprivation experiments, children with strabismic amblyopia have not been deprived of all vision in one eye since the first days of life.

If one eye is considerably more farsighted (hyperopic) than the other, then the weaker eye may be continually out of focus. Under these conditions, a condition called anisometropic amblyopia may develop in the more farsighted eye. Since humans do not have sharp vision at birth but develop 20/20 eyesight over their first three to five years, it is unlikely that this form of amblyopia is present in very early life.

Several scientists recognized that the monocular deprivation experiments were not a good model for most cases of human amblyopia. Anthony Movshon, Lynne Kiorpes, and their colleagues at New York University induced amblyopia in monkeys by less drastic means, such as putting a lens over one eye to reduce, but not completely eliminate, vision in that eye. Not surprisingly, these investigators found that the changes in circuitry in the visual cortex were not as severe as in the cortex of the monocularly deprived animals. For example, more connections survived between the compromised eye and the brain. Combined, these studies teach us that animal experiments are critical to understanding how the brain is wired, but we must be careful in generalizing from animal experiments to the human condition.

The most common approach to amblyopia is to treat only young children by patching the nonamblyopic eye. This treatment was used as early as 1743 by Georges-Louis Leclerc, the Comte de Buffon. Patching, or occlusion, of the "good" eye may strengthen connections between the amblyopic eye and the brain. This treatment is somewhat analogous to Hubel and Wiesel's experiments in which they closed one eye, then reopened it, and, at the same time, closed the other. With these manipulations, the input onto cortical neurons changed as well. Pharmacological treatments such as the use of atropine to blur vision in the "good" eye force the child to use the amblyopic eye and may produce effects similar to those resulting from patching. One drawback of occlusion thera- pies is that the open eye may improve its acuity at the expense of the occluded, or "good," eye. One eye gets better as the other gets worse.

What's more, patching may have adverse effects on binocular- ity and stereovision. Indeed, in his address upon winning the No- bel Prize for his research on vision, Torsten Wiesel expressed these concerns. A recent study by the federally funded Pediatric Eye Disease Investigator Group found that one-quarter of patients suc- cessfully treated with a patching regimen lost the acuity gains in the amblyopic eye one year after the patching regimen ended. The loss in eyesight may result because the patients still did not use their two eyes together. Something more or different was needed.

When Caitlin's mother was told that Caitlin would always have poor vision in her turned eye, she, like mothers everywhere, did not give up. She found Dr. Ruggiero instead. An amblyopic eye, Dr. Ruggiero told Caitlin and her mother, is amblyopic in all ways. Scientific studies have revealed that the amblyopic eye not only has reduced acuity but provides a distorted sense of space, tracks ob- jects less accurately, and is the poorer eye for directing hand move- ments. To reduce amblyopia, you must train the eye and brain in many different tasks. So, Caitlin practiced many exercises first with her amblyopic eye, then with her nonamblyopic eye, then with her amblyopic eye again. The number of tasks that a vision therapist can design to counter amblyopia are limited only by her imagination. For example, Caitlin practiced using one eye to pick up rice with tweezers, sort piles of colored rice into the different colors, or play number games like Yahtzee with miniature dice.

In 2007, Robert Hess and his colleagues at McGill University published a study demonstrating that people with amblyopia do not discount all information from the amblyopic eye. Instead, the amblyopic eye provides such weak input that it makes little contribution overall to what is seen. They suggested that therapy for amblyopia should include procedures in which the good eye is not patched but is put at a disadvantage. In this way, the input from the two eyes will be better balanced, and the patient may learn to use them more equally. In fact, this general strategy has been used by developmental/behavioral optometrists for the past forty years. They have developed a class of procedures called "monocular fixation in a binocular field" (MFBF). Dr. Ruggiero gave Caitlin several MFBF procedures to practice.

In MFBF procedures, both eyes are open, but only the amblyopic eye can be used for the difficult tasks. For example, Caitlin played the "wall game," a task I described in chapter 5. In this game, Caitlin looked at a set of red lights on a board. One light was illuminated at a time, and she had to find that light, hit it, then find the next illuminated light, hit it, and so on. Caitlin played this game while wearing red/green glasses. With the red lens over her amblyopic eye, only her amblyopic eye could see the illuminated light. So, Caitlin had to use her weak eye to find and hit the illuminated target, while the "good" eye, the eye behind the green lens, did not see the light but did see the general surroundings. This reversed the way Caitlin habitually saw since her amblyopic eye normally took a backseat to the better-seeing eye. Procedures like this helped Caitlin to balance and integrate the input from her two eyes. Once Caitlin had mastered these tasks, she was ready to practice binocular procedures, such as those described in earlier chapters.

Caitlin graduated from vision therapy after five months. It has been ten years, and Caitlin still maintains good vision in both eyes and has stereopsis. Her eyes are straight.

"How did you know," I asked Caitlin's mother, "if she was seeing differently after the vision therapy?"

Her mother told me that Caitlin stopped having trouble looking first at the words in her schoolbooks and then up at the blackboard. Caitlin used to grow restless with arts and crafts, but after
vision therapy, she protested if she had to stop coloring and go to
bed. She volunteered to do the reading when she and her mother
read aloud to each other. She no longer tilted her head to go up-
stairs. The head tilt had been a strategy to judge the height of the
steps, a habit she no longer needed once her stereovision improved. Best of all, Caitlin's confidence increased, especially with
dancing, her favorite activity.

"She is a child in motion," her mother said, "and now her motions are more fluid."

Caitlin was only six years old when she began vision therapy.
Are the same treatments effective in older children and adults?
When a person has amblyopia, do the connections from the amblyopic eye to the brain die out, or are these connections masked
by input from the "good" eye? In chapter 4, I described the case
of Red Greene, an amblyope who lost vision in his good eye in his
fifties. When this happened, eyesight in his amblyopic eye improved to 20/20. His case is not unique. In one published study,
one-third of the 203 amblyopes examined, many of whom were
older adults, experienced significantly improved eyesight in their
amblyopic eye after vision was lost in the fellow eye. Obviously,
loss of the "good" eye is not a treatment for amblyopia. However,
Red's story and this and other studies indicate that connections
from the amblyopic eye to the brain may survive throughout life
long after the close of the critical period. These connections, however, may be weak and suppressed.

Evidence lurking in the scientific literature for at least the last half
century indicates that amblyopia can be treated in adulthood. As
early as 1957, Carl Kupfer published a study in which he showed
dramatic improvements in adult amblyopes after a four-week period

of patching combined with vision therapy. In a 1977 study, Martin Birnbaum and his colleagues reviewed twenty-three published studies on amblyopia and reported that improvements in eyesight were found for all ages. Sadly, these and other studies continue to be overlooked. Many doctors still believe that amblyopia can be treated only in young children.

This is what Dr. Joan Lester, an anisometropic amblyope since early childhood, was repeatedly told. When she was sixty-five years old, however, a friend suggested that she consult a developmental optometrist. Dr. Lester saw Dr. Julie Kim, an associate at the office of Drs. Carole Hong and Kristina Stasko. Dr. Kim prescribed for Dr. Lester a program of optometric vision therapy that taught her how to use her long-neglected amblyopic left eye as an equal partner with her right. After six therapy sessions, Dr. Lester was ready to be fitted with a contact lens in the left eye. She described her first experience with the contact lens in this way: "Suddenly, the world sparkled. I felt an unaccountable joy explode. 'Oh my god,' I shrieked. . . . I felt unmoored, floating above the ground, dizzy." Dr. Lester described how each morning, when she slid the contact lens into the amblyopic eye, she experienced the same sudden "inexplicable joy." She felt like shouting, as if newly in love. Her friends say that Dr. Lester is now a changed person, more curious and eager for new adventure. At age sixty-seven, she has no plans to retire.

Scientific investigations, performed largely in the twenty-first century, give great credence to Dr. Lester's case. They suggest that the mature visual system is more plastic than once thought. For example, exposing adult rats to enriched environments reverses the amblyopia produced by monocular deprivation occurring in infancy. Placing adult rats in the dark for three to ten days can also reverse the effects of monocular deprivation. Studies of "perceptual learning" in human amblyopes pioneered by Dennis Levi and

Uri Polat in the mid-1990s demonstrate improvements in eyesight in the amblyopic eye even in adult patients. These studies and others demonstrate that visual acuity can be improved in the amblyopic eye of adults by training the participants to detect, for example, small discontinuities in a straight line or to discern faint lines against a slightly lighter background. In contrast to the results with patching, but similar to the results with vision therapy, eyesight improved in the amblyopic eye without a loss of vision in the "good" eye.

———

While the best approach may vary from patient to patient, one basic principle needs to change. The brain may be more plastic, more responsive to treatments in infancy, but this period of high malleability does not exclude the possibility that improvements can occur later in life. If an effective treatment—be it surgery, eyeglasses, occlusion, or optometric vision therapy—is available, it should be applied at the youngest age possible. For example, surgery for infantile strabismus may be most effective if it straightens the eyes in a stable manner within three months of the initial eye misalignment. But a period of three months from strabismus onset to surgery is a very short window, and many patients will miss this period.

Only 20 percent of patients with infantile strabismus who have surgery after the age of two years develop stereopsis. In contrast, 38 to 50 percent of patients with infantile strabismus and 70 percent of patients whose strabismus develops after the first year acquire stereopsis through optometric vision therapy. Treatments should not be denied to older children and adults solely on the basis of age.

I had lectured in the classroom for seventeen years on early critical periods in visual development before I discovered that I was not permanently stereoblind. When I first went to see Dr. Ruggiero, I did not believe I could gain stereovision but was hoping instead to find a way to stabilize my gaze. No wonder gaining stereopsis was one of the most powerful and liberating experiences of my life. But now I had a new and important question. What specific conditions facilitate changes in the adult brain? Could my experience apply to rehabilitation in general?

9

VISION AND REVISION

Time for you and time for me,
And time yet for a hundred indecisions,
And for a hundred visions and revisions,
Before the taking of a toast and tea.

—From "The Love Song of J. Alfred Prufrock," by T. S. Eliot

Four years after I learned to see in stereo, I visited Dr. Ruggiero for my annual exam. I had graduated from weekly vision therapy sessions three years earlier, although I still practiced some vision therapy techniques at home every day. At the time of this yearly checkup, I was caring for my ill father. Since I had no idea that my mental state could influence how I used my eyes, I didn't mention my personal difficulties to Dr. Ruggiero. She picked up her occluder and moved it in front of my eyes in order to complete a quick test of eye alignment. With the sight of the occluder, I felt myself slip back in time, back to being the little cross-eyed girl in my eye surgeon's office. The doctor was waving the occluder in front of my face and asking me to focus at a point a short distance away. I felt naked and exposed, powerless and confused.

Dr. Ruggiero moved the occluder in front of one eye then the other. "What is this?" she asked. "Your left eye turned in and just sat by your nose. I haven't seen this behavior since you started vision therapy." I told her that thoughts of my father and the standard eye alignment test had taken me back to my childhood. Could she do the test again? Dr. Ruggiero repeated the test, and with a determined effort, I brought myself back to the present. This time, my eyes remained straight.

This experience both intrigued and frightened me. I learned that I actually had two ways of seeing. I saw with stereopsis almost all the time, but I could revert to my old strabismic ways under stressful conditions. My experiences with vision therapy had already taught me that we can't understand vision without making connections between sight, spatial orientation, and movement. Now I discovered that we can't understand how we perceive the world and how we adapt and learn without considering the whole person—the thinking, moving, and feeling person.

Past studies of the nervous system, like those mentioned in the previous chapter, have often involved recordings from the brains of anesthetized animals. If an animal is anesthetized, then we are not studying its brain activity in a normal behavioral state. While these experiments have provided us with a tremendous amount of information on nerve circuits, they can only partially explain how an animal can learn and change. Add an actively engaged animal or an attentive and motivated human being to the picture, and a new story emerges.

In his fascinating book *Rebuilt: How Becoming Part Computer Made Me More Human,* Michael Chorost, who was hard of hearing as a young child and profoundly deaf in adulthood, describes how he learned to hear again using a cochlear implant. When Chorost first heard through his cochlear implant, he perceived lots of beeps and whistles and could hardly make sense of the sounds. Yet, Chorost, like many other deaf individuals, learned to use his implant to understand human speech—even over a telephone. He describes how he rehabilitated his own hearing and states that this accomplishment, unlike hearing in a normal baby, was an intensely conscious act.

Changing our vision was also an intensely conscious act for me and the others I have described. A child (whose behaviors are less entrenched than an adult's) can make great strides with vision therapy simply by following instructions, but adults receive the most benefit if they think carefully about what they are doing and why. It takes great awareness and concentration to change your habits and the way you've always negotiated the world. Luckily, this level of concentration is usually not required forever. Once we learn a more efficient, more informative way of approaching the world, new habits replace the old.

What a magnificent feeling it is to take control of your own vision and solve your own problems. I will never forget the epiphany that I had with the Brock string—the glorious feeling of aiming my two eyes toward the same point at the same time. It was incredibly liberating to discover that I could tackle a visual problem that had hounded me since childhood. I could transform my own vision.

What is true for me is generally true for people who recover from injuries and other physical problems. Edward Taub and his colleagues at the University of Alabama discovered that the patients most actively involved in their own rehabilitation recovered best from strokes. These were the patients who planned out how they would use the weakened limb in daily tasks, kept a diary of their actions, and solved problems as they came up. Taub's patients had to heal themselves.

When I began vision therapy, I thought the procedures seemed too straightforward, simple, and low-tech to be effective. But just the opposite was true. If my therapy had involved stereo tasks performed only with complicated instruments and stereoscopes, I might not have generalized these skills to daily life. Instead, I practiced aiming my eyes at actual objects in real space while moving in routine ways. Indeed, as early as the mid-1900s, Frederick Brock stressed these ideas when he noted that to train strabismics to see with normal binocular vision and stereopsis, you have to challenge them with tasks that are close to, but just beyond, their current skills and with exercises that resemble actions experienced in real life. Recent scientific studies on learning reveal that Frederick Brock was clearly ahead of his time.

Twenty-first-century research indicates that the adult nervous system can change its circuitry, but it rewires only in response to behaviorally important stimuli. In contrast, an infant nervous system may change its connections in response to any stimulus as long as it is very strong or repeated sufficiently. For example, barn owls use both their eyes and ears to locate and catch a small mouse on the forest floor. If a scientist places prisms over the eyes of the owl, the owl's senses are disrupted. Its eyes tell it the mouse is in one location, while its ears tell it the mouse is in another. Young barn owls automatically change the wiring in their brains to deal with the visual-auditory mismatch. This happens whether or not the young owls are caged and fed or forced to hunt for their food. Adult barn owls who are caged and fed do not adjust. If, however, these adults are forced to hunt, then they realign their brain's auditory spatial maps with the prism-altered visual maps. Neuronal connections in their brain change so that they can catch their food, clearly a behaviorally important task.

My own visual experiences, the rehabilitation accounts of other people, such as those recovering from a stroke, perceptual learning studies, and recent animal studies suggest a new definition for the critical period. The critical period encompasses the developmental stage when the brain changes in response to most strong stimuli, not just to behaviorally relevant ones. A very young nervous system responds passively to whatever stimuli are present. As we grow older, we become more selective. Rewiring in the adult brain requires active learning. Now scientists are discovering which areas of the brain are involved in this process and how they may influence our nervous system.

To learn to see in 3D, I needed to engage, from an evolutionary point of view, both the newer and older parts of my brain. As I suggest in chapter 6, some of the changes in my brain involved modifications of synaptic connections in the visual cortex, an area of the brain that is highly evolved in humans but much less developed in our distant vertebrate cousins. The visual cortex is part of the cerebral cortex, a brain structure that has grown so large in people that it wraps itself around almost the entire surface of the brain. In addition to the visual cortex, located at the back of the brain, the cerebral cortex includes several other sensory areas, such as the auditory cortex located on the sides and the somatosensory (body sense) cortex located at the top of the brain. Neuronal connections in many areas of the cerebral cortex, including the sensory areas, change with experience. In fact, the wiring in the cerebral cortex reflects the history of our actions and is constantly reshaped by them.

But not all experiences provoke long-lasting changes in cortical circuitry. Some experiences are more powerful than others. Areas of our brain located outside the cerebral cortex are very important in assigning meaning and significance to various experiences and then stimulating the connections in the cortex to change. These noncortical areas are brain regions that we share with all our vertebrate cousins, including fish, amphibians, reptiles, birds, and mammals. Such areas include parts of the brainstem at the base of the brain as well as an area toward the front of the brain called the basal forebrain. The brainstem and basal forebrain contain neurons that send their axons all over the central nervous system. At their axon tips, the neurons secrete chemicals called neuromodulators, which include noradrenaline (also called norepinephrine), serotonin, dopamine, and acetylcholine. By delivering these neuromodulators throughout the nervous system, these ancient parts of the brain have widespread effects on basic behaviors and behavioral states, such as sleep-wake cycles, general levels of arousal, overall mood, learning, and memory.

Activation of these neuromodulatory areas has striking effects on the circuitry of the cerebral cortex. For example, stimulation of the basal forebrain can change the way the auditory cortex responds to sound. In several different experiments, adult guinea pigs or rats were played a sound of a particular pitch, while neuronal activity was monitored in their auditory cortex. Simple repetition of the tone did not change the response of adult auditory cortical neurons. The same proportion of neurons continued to respond to the particular tone as before the experiments. Electrodes were then used to stimulate the basal forebrain at the same time that the animals heard the tone. Under these conditions, the auditory cortex reorganized. Now more neurons were responsive to the particular tone. So, simultaneous activity of the basal forebrain with the presentation of the tone produced a significant change in the wiring of the adult auditory cortex. Similar results have also been seen when presentation of a particular stimulus is paired with activation of other neuromodulatory areas of the brain. Activation of these areas releases neuromodulators onto cortical neurons that trigger or facilitate changes in cortical circuitry.

When I learned to aim both eyes simultaneously at the same point in space, synaptic connections from both eyes were strengthened onto individual visual cortical cells. Could the release of neuromodulators onto my cortical neurons have facilitated these synaptic changes? The neuromodulators may have helped to unmask and strengthen connections that had been ineffective but were never entirely lost. In addition, the same neuromodulators that triggered and facilitated synaptic changes may have made these changes long-lasting.

If neuromodulators were involved, then I must have engaged in therapy activities that stimulated the neuromodulatory areas of the brain. High levels of neuronal activity in the brainstem and basal forebrain are seen in animals when they are alert and exploring their environment, when they are learning about novel stimuli, and when they anticipate a reward for their actions. When I first read about the conditions under which neuromodulatory areas are active, I immediately thought back to my first experiences with the Brock string. With that procedure, I experienced new sensations both in the way I felt my eyes move and in a novel perception of depth. I remember exactly where I was standing and how I told the therapist, somewhat breathlessly, that I could feel my eyes move. What's more, I felt triumphant. Accomplishing the task provided a great reward. All of these reactions would powerfully activate the neuromodulatory areas of my brain.

Heather Fitzpatrick experienced something quite similar the first time she "got" the Brock string. Cross-eyed from the age of two, she described to me what it felt like to see the image of the strings form an X around the bead, to be aiming her eyes at the same place at the same time.

> "I love that X!!!! I like the way my head feels once I get it. . . . [I]t is a bit like finding the perfect arm position when swimming, you get the maximum pull with the least effort. . . . It feels as though things are in sync. I actually feel my brain looking through my eyes and it is this perfect balance . . . !"

Had a scientist been recording neuronal activity from Heather's brain at the moment she saw the X, he probably would have seen the neuromodulatory areas of her brain firing like mad, liberating their chemical messengers onto her visual cortical cells and facilitating learning changes in her brain.

When my vision began to change, all I wanted was to be left alone to look and explore. I spent hours simply looking at ordinary objects like straws sticking out of plastic drink glasses, building pipes suspended from the ceiling, and telephone poles in long, straight rows. My surprising and delightful views sent me on hunts for new sights. I loved going to greenhouses where the plants popped out at me or modern art galleries where I could examine the intriguing pockets of space within each piece of sculpture. I was in a constant exploratory state, activating the neuromodulatory areas of my brain. I wonder if babies' brains are so malleable in part because everything is new to them. Their neuromodulatory areas might be continually active.

The unmasking or revival of previously weak synapses may explain recovery in many other situations as well, such as in the somatosensory cortex after a stroke. Different neurons in the somatosensory cortex respond to touch from different parts of the body. If somatosensory neurons that receive input from your fingers are damaged, then you will have trouble sensing and using those fingers. However, intense practice of a task specifically requiring the use of the fingers may lead to functional recovery, allowing you to regain the use of your hand.

Scientists cannot easily record from human brains to determine exactly where recovery has occurred. We can, however, record from the brains of monkeys who have suffered strokes, then recovered function in their affected fingers through exercises specifically requiring the use of them. After training, neurons in the uninjured areas of the somatosensory cortex respond to touch from the affected fingers. Many of these neurons may have received very weak connections from these finger pathways all along. Only after the injury and rehabilitation, however, did those synapses become effective.

If we think about the ways we sense the world, then connections, some strong and some perhaps hidden, should exist among the parts of the brain that respond to touch, sight, and hearing. We rarely experience our surroundings through vision alone but instead develop a coherent perception of the world through a combination of our senses. Not surprisingly, then, neurons have been found in the brain that are sensitive to both tactile and visual input and are involved in our ability to reach for and grasp objects. The responses of some neurons in the visual cortex are modulated by sounds. If we lose our sight, the other senses can use parts of the brain once devoted mostly to vision. Under these conditions, hidden connections between different sensory areas of the brain may be uncovered.

The books *Privileged Hands* and *Touching the Rock* both highlight in fascinating detail what it is like to be blind. In a beautiful passage in *Touching the Rock,* John Hull describes the unique sounds that rain makes when it falls on different surfaces in his yard and how the sounds help him to create a global sense of his surroundings. Both authors make use of echoes in ways that sighted people rarely, if ever, do. When they walk, their footsteps create sounds that bounce off the objects around them, and they use the direction of these echoes to navigate. Many of these skills may result from the use of connections in the brain that normally lie untapped in a sighted person but are uncovered when sight is lost.

Indeed, imaging studies of the brains of individuals who are born blind or become blind in early life indicate that they use their visual cortex for nonvisual activities. To read braille, for example, you use your index finger to feel the bumps on the page that represent the letters. Not surprisingly, the somatosensory cortex is activated when sighted or blind individuals read braille. What is surprising is that the visual cortex also lights up in blind people's brains but not in the brains of people who can see. In *Privileged*

Hands, Geerat Vermeij describes how he studies the evolution of mollusks by using his exquisite sense of touch to analyze their shells. His visual cortex is probably active when he does so. There are connections between the visual cortex and the parts of the brain that mediate touch and hearing, and these connections, silent and ineffective in sighted people, are active in the blind.

Since a blind individual negotiates the world very differently, a seeing person cannot know what it is like to be blind by briefly closing his eyes. In fact, this situation has created some misunderstandings in designing technology for the blind, such as the creation of audible traffic signals. By producing signals that would buzz or beep when it was safe to cross the street, engineers thought they would be helping blind people negotiate traffic. However, the National Federation of the Blind has reported that audible traffic signals can actually compromise safety. By beeping and buzzing, the signals create extra noise and confusing echoes, making it hard for a blind person to hear and locate oncoming cars. While the traffic signals were designed with the best of intentions, the designers may have assumed that a blind person is an individual who simply cannot see. Instead, someone who is blind has developed a brain and skills to navigate through the world without vision. He or she may exploit different sets of connections in the brain.

In Spain, future teachers of the blind must spend a week with blindfolds on in an attempt to experience firsthand what it is like not to see. By the end of the week, the instructors have noticed that they are better able to recognize people by the sound of their voice or the cadence of their walk. Some are even able to detect obstacles in their path by hearing the echoes of their own footsteps as they bounce off objects.

Alvaro Pascual-Leone and his colleagues at Harvard University blindfolded sighted individuals for one week and imaged their brains while the participants attempted to identify braille letters or the pitch of a sound. At first, the visual cortex was silent during these nonvisual tasks. After five days, however, areas of the visual cortex became active. The scientists postulated that after only five days of blindness, input from the touch and auditory systems activated the visual cortex via connections that were always present but normally weak or silent. When sight was restored, these connections were masked once again.

Investigations such as these make me wonder about the use of the term *hardwired* when we talk about the human brain. How many connections are truly hardwired, and how many can change under different circumstances? It was once thought that the visual cortex was hardwired only for sight, but studies on the blind and blindfolded indicate that this is not the case. We may have many connections in our brain that remain untapped unless circumstances, from learning a new skill to recovering from a stroke, require their participation. The critical factor may be in designing therapeutic conditions that unmask these synapses. For me, vision therapy provided the conditions that allowed me to revive synapses onto visual cortical neurons that had remained weak and ineffective throughout most of my life. When I consider the extraordinary changes produced in the blindfold study described above, the alterations that may have occurred in my brain do not seem so surprising.

Our conventional and limited view of adult neuronal plasticity derives in part from the specific ways that scientists and physicians have designed laboratory experiments and clinical therapies. We cannot understand neuronal plasticity by studying brain circuits in isolation from the whole person. Only by considering a person's adaptations and response to her condition can we really explore the amazing plasticity of the human brain to rewire itself throughout life in order to recover from injury, learn new skills, improve perception, and even gain new qualia.

———

On February 12, 2008, six years to the day after I received my first prism glasses, I went into Dr. Ruggiero's office to pick up a new pair of spectacles. A year earlier, Dr. Ruggiero had cut my prism correction in half. With this new pair of glasses, I had no prism correction at all. When I put on the glasses and stood up, I felt disoriented, but the feeling lasted for only a moment. I drove back to work wearing my new glasses, very aware of how I was seeing. I could take in the scene all around me. I felt like I was moving within a fixed and stable world. Branches on the trees by the road appeared in layers of depth, and the mountains in the near distance stretched not just up but out toward the horizon.

Later that evening, I tested myself on every vision therapy procedure I had learned during my stereo journey. I could still aim my eyes together along the entire length of the Brock string. I could still fuse images in the quoits vectogram and see the rope circle float in space. I had feared that without the prism, I would go back to my cross-eyed ways. But I needn't have worried.

I finally relaxed and thought back on all that had happened. When I first learned about stereovision in college, I assumed that if I could see in 3D, I would be better at threading a needle, parking a car, and hitting a tennis ball. Of course, all of this is true, but I had no idea just how different and how magnificent the world would appear in all its glorious dimensions. Most importantly, I learned that I was not the victim of a visual fate that had been sealed in early childhood. I could rehabilitate my own vision. My newfound and hard-earned stereovision has given me an enormous sense of security, confidence, and accomplishment. It is with a stable, clear, and depth-filled gaze that I now encounter the world.

Acknowledgments

I would not have written this book without the encouragement and guidance of two exceptional individuals. I thank my optometrist, Theresa Ruggiero, for her expert clinical skills, for the hours that she spent discussing her work with me, and, most of all, for teaching me that I could rehabilitate my own vision.

I owe a great debt to Oliver Sacks, who read and responded to my many letters and also wrote "Stereo Sue" for the *New Yorker* and the foreword for this book. I am particularly thankful to Dr. Sacks for encouraging me to seek out and consider the stories of other individuals with visual experiences similar to my own. It was Theresa Ruggiero who taught me how to see and Oliver Sacks who taught me how to listen.

I thank Dr. Ruggiero's vision therapists, Michelle Dilts, Ellen Middleton, and Laurie Sadowski, for their guidance and humor in the vision therapy room, and I am very grateful to optometrist Steven Markow for referring me to Dr. Ruggiero.

I thank the many people who generously and enthusiastically shared their own vision stories with me, including Bruce Alvarez, Garry Brown, Stephanie Willen Brown, Eliza Cole, Rachel Cooper, Margaret Corbit, Jennifer Clark Deegan, Eric Dore, Michelle Dore, Andre Dore, Pat Duffy, Heather Fitzpatrick, Tara Fitzpatrick, Tracy Gray, Richard (Red) Greene, Rachel Hochman, Joan Lester,

Margaret Lundin, Sarah Merhar, Cyndi Monter, Rebecca Penneys, Steve Perez, Caitlin Ronayne, Michelle Scully, Lucas Scully, Kathie Terry, Oliver Waldman, Eric Woznysmith, and Richard Zack.

Optometrists Amiel Francke, Israel Greenwald, Carl Gruning, Paul Harris, Hans Lessmann, and Leonard Press spent hours discussing their work with me and allowed me to visit their practices. And special thanks to Leonard Press and Paul Harris for sharing with me their encyclopedic knowledge of vision and for responding to many e-mails—too numerous to count.

Several people read and offered helpful comments on all or substantial portions of this book, including Benjamin Backus, Nigel Daw, David Hubel, Leonard Press, Gerald Westheimer, and Jeremy Wilmer. I am particularly grateful to David Hubel for his support and guidance.

I thank my former professors, Allan Berlind and Alan Gelperin, for introducing me to the fields of neuroscience and binocular vision, and Alan Gilchrist for introducing Dr. Ruggiero to visual psychophysics. Though I will inevitably omit many others who should be mentioned, I am grateful to the following colleagues, scientists, optometrists, ophthalmologists, and friends: Jacob Bloomberg, Michael Chorost, Ken Ciuffreda, David Cook, Frank Durgin, Kate Edgar, Barbara Ehrlich, Manfred Fahle, Rachel Fink, Nathan Flax, Teresa Garland, Ray Gottlieb, Carl Hillier, Carole Hong, Claire Hopley, Jonathan Horton, William Huebner, Carolyn Hurst, Eric Knudsen, Nancy Lech, Dennis Levi, Robin Lewis, Margaret Livingstone, Diana Ludlam, John Merritt, Donald Mitchell, Maureen Powers, Bill Quillian, Stan Rachootin, Joann Robin, Shinsuke Shimojo, Ralph Siegel, Susan Smith, Elizabeth Socolow, John Streff, Diana Stein, Cathy Stern, Selwyn Super, Barry Tannen, Nancy Torgerson, Christopher Tyler, Bob Wasserman, and Qasim Zaidi. I also benefited from discussions with my students Heather Ajzenman, Cordelia Erickson-Davis, and Katie Wagner.

I am grateful to Margaret Nelson and my student Julia Wagner for the excellent illustrations in this book and Rosalie Winard for superb photographs. Thanks also to James Gehrt, Mary Glackin, and Wendy Watson at Mount Holyoke College for help in the preparation of illustrations.

Anne Drury and Ajay Menon of the Mount Holyoke College Library were able to retrieve even obscure scientific papers for me through interlibrary loan. Thanks also to Joan Grenier and Bob L'Episcopo of the Odyssey Bookshop for help in obtaining rare books and Rachel Cooper for her highly informative websites.

I am grateful to Robert Krulwich of National Public Radio, who aired a *Morning Edition* program on my vision story entitled "Going Binocular: Susan's First Snowfall."

Financial support for writing this book was provided through a faculty grant from Mount Holyoke College.

I owe enormous thanks to my book agent, Lisa Adams of the Garamond Agency, who guided me through the process of drafting a proposal and the business of publishing a book and to Amanda Moon, my editor, and Jen Kelland, my copy editor, at Basic Books, who helped me turn a collection of stories and ideas into a coherent narrative. I thank Whitney Casser and Collin Tracy for publishing support.

I am deeply grateful to my family: to my artist father, Malcolm Feinstein, for allowing me to reproduce some of his paintings in this book; to my children, Jenny and Andy Barry, for carving out a room in our house to be my book-writing study; and to my swashbuckling husband, Dan Barry, for his unflagging support and incorrigible optimism.

Finally, I would like to thank a man whom I will never meet, the late Frederick W. Brock (1899–1972). His daughter, Dolores Brock Partridge, and grandson-in-law, Bruce Alvarez, provided me with all of Frederick Brock's writings. Israel Greenwald, Brock's optometric partner, spent hours describing Brock's techniques to me. Frederick Brock's articles on strabismus and its treatment have remained largely unread in the optometric literature for too long. I hope with this book to bring his research and insights into the light of day.

Glossary

Accommodative esotropia—a type of strabismus that often develops at about two to three years of age. If a child is farsighted, then he must focus the eyes (accommodate) with extra effort to see a near target clearly. He also turns in, or converges, the eyes so that the image of the target is cast on the foveas of both retinas. Since the processes of accommodation and convergence are linked, the extra effort needed to focus may result in overconvergence of the eyes. If this overconvergence becomes habitual, accommodative esotropia develops.

Accommodative/vergence disorders—a set of visual disorders in which there is poor coupling between aiming the two eyes at the target (vergence) and focusing the eyes on the target (accommodation). Convergence insufficiency is a type of accommodative/vergence disorder.

Alternating esotropia—a condition in which the individual alternates between the eye he uses to fixate the target and the eye he turns in.

Amblyopia—loss of vision in one or both eyes that cannot be corrected by lenses and is not due to eye disease. This condition is colloquially termed *lazy eye* or *wandering eye*. Amblyopia is present in 2 to 3 percent of the population and is the leading cause of monocular blindness in the under-age-forty population.

Anisometropia—the condition in which the two eyes have different refractive powers. One eye may be more nearsighted (myopic) or farsighted (hyperopic) than the other. If one eye is more farsighted, then it will be continually out of focus. This condition can lead to amblyopia in the more farsighted eye.

Anomalous correspondence—a condition found in strabismics in which objects seen by the foveas of the fixating and the turned eyes are not interpreted as being located at the same point in space. Instead, objects seen by the fovea of the fixating eye and a nonfoveal region of the misaligned eye are interpreted as being in the same place. In many cases, the images seen by the two foveas are interpreted as being separated in space by the same amount as the angular deviation of the two eyes.

Behavioral optometrist (developmental optometrist)—an optometrist with a specialty in treating functional vision problems, including difficulties with binocular vision and depth perception, eye movements, visual problems that impact reading and learning, and visual deficits following stroke or brain injuries. These optometrists are skilled in the use of optometric vision therapy.

Bifixate—to aim the fovea of both eyes at a target.

Convergence—the turning in of the two eyes when one looks at a near distance. You converge your eyes to position the image of regard onto the fovea of both retinas.

Convergence insufficiency—a disorder in which the individual does not focus and aim the eyes at the same place in space, particularly when viewing a near target.

Corresponding retinal regions—the fovea of each retina as well as the regions located equally to the right or equally to the left of them. For people with normal binocular vision, objects that cast their images on corresponding retinal regions are interpreted as being located in the same place in space.

Crossed eyes—a colloquial term for esotropia.

Depth perception—the sense of an object's distance obtained from both stereopsis and monocular depth cues like perspective, shading, and object occlusion.

Developmental optometrist (behavioral optometrist)—an optometrist with a specialty in treating functional vision problems, including difficulties with binocular vision and depth perception, eye movements, visual problems that impact reading and learning, and visual deficits following stroke or brain injuries. These optometrists are skilled in the use of optometric vision therapy.

Divergence—the turning out of the two eyes when switching one's gaze from near to far. You diverge the eyes to position the image of regard onto the fovea of each retina.

Esotropia—a condition in which one eye fixates, or is aimed at, the target, while the other eye turns in. This disorder is colloquially called crossed eyes.

Exotropia—the condition in which one eye fixates, or is aimed at, the target, while the other eye turns out. This condition is colloquially called walleye.

Fixate—to aim the eye at a target so that the image of the target is cast on the retinal fovea.

Fovea—the central region of the retina containing the highest density of light-sensing cells. Images are seen most clearly and in the greatest detail if cast on the fovea.

Infantile esotropia—esotropia that develops within the first six months of life. This condition is also called congenital esotropia, although it is rare for an infant to be born with esotropia.

Lazy eye—a colloquial term for amblyopia.

Normal correspondence—the condition found in normal vision in which objects seen by the two foveas are interpreted as being located in the same place in space.

Panum's fusional area—the area on the two retinas that corresponds roughly to the plane where one is looking. Objects whose images fall within Panum's fusional area are seen as single. Objects whose images fall outside of Panum's fusional area are seen as double. See discussion of the Brock string in chapter 6.

Retina—the tissue located toward the back of the eye that contains the light-sensing rod and cone cells.

Squint—another term for strabismus.

Stereopsis—the ability to use the different viewing perspectives of the two eyes to see in three dimensions.

Strabismus—a misalignment of the visual axes of the two eyes. Esotropia (crossed eyes) and exotropia (walleye) are both types of strabismus.

Walleye—a colloquial term for exotropia.

Wandering eye—a colloquial term for amblyopia.

Resources

College of Optometrists in Vision Development (COVD)

COVD is a nonprofit, international organization whose members include optometrists, optometry students, and vision therapists. The organization was founded in 1971 to standardize optometric vision therapy techniques and provide a rigorous board-certification process for optometrists and vision therapists. Optometrists who complete the board-certification process become fellows of COVD. The COVD publishes the journal *Optometry and Vision Development*. Information about the COVD can be accessed via their website at www.covd.org.

Optometric Extension Program Foundation (OEP)

Optometrists A. M. Skeffington and E. B. Alexander founded the Optometric Extension Program Foundation in 1928 to provide educational programs and materials for both eye-care professionals and the public. The organization publishes the *Journal of Behavioral Optometry*. Information about the OEP and useful books about vision therapy can be accessed via their website at www.oepf.org.

Neuro-Optometric Rehabilitation Association (NORA)

NORA includes a group of diverse professionals interested in the rehabilitation of individuals suffering from visual-motor, visual-perceptual, and visual-information-processing dysfunctions. More information can be accessed via their website at www.nora.cc.

Find a Doctor

The above three organizations (COVD, OEP, and NORA) provide directories for optometrists conversant in optometric vision therapy.

InfantSEE

InfantSEE is a public health program designed to provide comprehensive vision assessments to children in the first year of life. Optometrist members of the InfantSEE program provide vision evaluations free of charge as a public health service. More information can be obtained via their website at www.infantsee.org.

Useful Websites

Optometrists Network provides a wealth of information at the following linked websites:

optometrists.org
strabismus.org
lazyeye.org
convergenceinsufficiency.org
children-special-needs.org
braininjuries.org
learningdisabilities.net
vision3d.com
visionstories.com
visiontherapy.com, visiontherapy.org

These highly informative and recommended websites contain no ads and were created by Rachel Cooper, an individual who suffered from poor binocular vision until she received optometric vision therapy in her mid-thirties.

Parent Organizations

Parents Active for Vision Education (PAVE)

Founded by concerned parents, PAVE is dedicated to helping children who struggle in school as a result of visual disorders. More information can be accessed at their website at www.pavevision.org.

Vision First Foundation

The Vision First Foundation was founded by Janet Hughes whose mission is to make others aware that school vision screenings are not sufficient to uncover and diagnose many vision problems that impede a child's progress through school. Ms. Hughes was active in the passage of an Illinois law requiring written notification before a vision screening that states, "Vision screening is not a substitute for a complete eye and vision evaluation by an eye doctor." More information can be obtained via the foundation's website at www.visionfirstfoundation.org.

Books

Numerous books about vision therapy are available via the OEP website at oep.excerpo.com. COVD fellow optometrist David Cook has published two helpful books concerning vision and vision therapy: *Visual Fitness* (2004) and *When Your Child Struggles* (1992).

Notes

Chapter 1: Stereoblind

1   *He was describing the development of the visual system, highlighting experiments done on walleyed and cross-eyed kittens.*

Hubel DH, Wiesel TN. Binocular interaction in striate cortex of kittens reared with artificial squint. *Journal of Neurophysiology* 28 (1965): 1041–59.

Hubel DH, Wiesel TN. *Brain and Visual Perception: The Story of a 25-Year Collaboration.* Oxford: Oxford University Press, 2005.

2   *We have two eyes, he said, but only one view of the world.*

An excellent review of stereovision is found in Hubel DH. *Eye, Brain, and Vision.* New York: Scientific American Library, 1995. Also available online at http://hubel.med.harvard.edu/bcontex.htm.

3   *The classroom didn't seem entirely flat to me.*

See chapter 7 for a more detailed discussion of nonstereoscopic cues to depth.

4   *Many of the great students of optics, including Euclid, Archimedes, da Vinci, Newton, and Goethe, never figured out how we see in stereoscopic depth.*

For a fascinating review of the history of our knowledge of binocular vision, see

Howard IP, Rogers BJ. *Seeing in Depth.* Ontario: I. Porteus, 2002, vol. 1, ch. 2.

Crone RA. *Seeing Space.* Exton, PA: Swets and Zeitlinger Pubs., 2003.

6   *The teddy bear located to your left casts its image on corresponding points on the right side of both your retinas, while the rattle, to the right, casts its image on corresponding points on the left side of both retinas.*
      In figure 1.3, the block is located at the fixation point. The bear, block, and rattle all fall on a line called the horopter, or, more specifically, the apparent frontoparallel plane horopter. Although this line is slightly curved, we interpret the bear, the rattle, and all other objects along this horopter as located in the same plane as the block. See, for example,
      Steinman SB, Steinman BA, Garzia FP. *Foundations of Binocular Vision: A Clinical Perspective.* New York: McGraw-Hill Cos., 2000, ch. 4.

6   *In 1838, Wheatstone explained how the relative position of the images on the two retinas allows us to see in 3D.*
      Wheatstone C. Contributions to the Physiology of Vision.—Part the First. On some remarkable, and hitherto unobserved, Phenomena of Binocular Vision. *Philosophical Transactions of the Royal Society of London* 128 (1838): 371–94.

10   *I thought about people who were totally colorblind.*
      Individuals with a hereditary form of total colorblindness are described in
      Sacks O. *The Island of the Colorblind.* New York: Alfred A. Knopf, 1996.

10   *With this knowledge, could they see in their mind's eye what they couldn't see in the real world?*
      Many philosophers have pondered this question. For example, Frank Jackson posed the famous thought experiment about Mary the Neuroscientist. Mary is brilliant and knows everything there is to know theoretically about color and color vision. However, she has lived all her life in a black-and-white room, her entire body covered in black-and-white clothes, so that she saw absolutely no color. Finally, Mary is let out of her room. She sees red for the first time. Is red what she imagined? Had she been able to imagine any of the colors that she now sees? Has she learned something new about the world?
      Jackson F. What Mary didn't know. *Journal of Philosophy* 83 (1986): 291–95.
      Also see Ludlow P, Nagasawa Y, Stoljar D (eds.). *There's Something about Mary: Essays on Phenomenal Consciousness and Frank Jackson's Knowledge Argument.* Cambridge, MA: MIT Press, 2004.
      Thanks to Greg Frost-Arnold for introducing me to Mary.

11   *If you learn to read braille even as an adult, the number of neurons in your brain that receive touch input from your reading index finger increases.*

Pascual-Leone A, Torres F. Plasticity of the sensorimotor cortex representation of the reading finger of braille readers. *Brain* 116 (1993): 39–52.

11   *In the 1990s, scientists studied the brains of violinists with magnetic source imaging, and they found that more neurons in the motor cortex of violinists were devoted to the control of the fingers of the left than the right hand.*

Elbert T, Pantev C, Wienbruch C, Rockstroh B, Taub E. Increased cortical representation of the fingers of the left hand in string players. *Science* 270 (1995): 305–7.

13   *You might think that Jenny's experiments indicated that Dan's ability to sense and move had degenerated while he was in space. Instead, Dan had adapted to a radically new environment, the microgravity of outer space.*

Reschke MF, Bloomberg JJ, Harm DL, Paloski WH. Space flight and neurovestibular adaptation. *Journal of Clinical Pharmacology* 34 (1994): 609–17.

Reschke MF, Bloomberg JJ, Harm DL, Paloski WH, Layne C, McDonald V. Posture, locomotion, spatial orientation, and motion sickness as a function of space flight. *Brain Research Reviews* 28 (1998): 102–17.

14   *Since my misaligned eyes saw different things, they competed for input onto visual cortical neurons, and on each neuron, one or the other eye won out.*

Hubel DH, Wiesel TN. Binocular interaction in striate cortex of kittens reared with artificial squint. *Journal of Neurophysiology* 28 (1965): 1041–59.

Wiesel TN, Hubel DH. Comparison of the effects of unilateral and bilateral eye closure and cortical unit responses in kittens. *Journal of Neurophysiology* 28 (1965): 1029–40.

15   *But even at the turn of twenty-first century, the latest papers and books, though full of evidence of the adaptability of the adult brain, still didn't question the critical period in relation to stereovision.*

See, for example, Gilbert CD, Sigman M, Crist RE. The neural basis of perceptual learning. *Neuron* 31 (2001): 681–97.

Begley S. *Train Your Mind, Change Your Brain: How a New Science Reveals Our Extraordinary Potential to Transform Ourselves.* New York: Ballantine Books, 2007, 77–78.

In contrast, other papers and books, largely written after 2002, suggest that substantial plasticity may extend beyond the critical period.

Bao S, Chang EF, Davis JD, Gobeske KT, Merzenich MM. Progressive degradation and subsequent refinement of acoustic representations in the adult auditory cortex. *Journal of Neuroscience* 23 (2003): 10765–75.

Doidge N. *The Brain That Changes Itself.* London: Penguin Books, 2007.

Fahle M, Poggio T. *Perceptual Learning.* Cambridge, MA: MIT Press, 2002.

Kasamatsu T, Watabe K, Heggelund P, Scholler E. Plasticity in cat visual cortex restored by electrical stimulation of the locus coeruleus. *Neuroscience Research* 2 (1985): 365–86.

Keuroghlian AS, Knudsen ET. Adaptive auditory plasticity in developing and adult animals. *Progress in Neurobiology* 82 (2007): 109–21.

Levi DM. Perceptual learning in adults with amblyopia: A reevaluation of critical periods in human vision. *Developmental Psychobiology* 46 (2005): 222–32.

Li RW, Klein SA, Levi DM. Prolonged perceptual learning of positional acuity in adult amblyopia: Perceptual template retuning dynamics. *Journal of Neuroscience* 28 (2008): 14223–29.

Reports over the past century indicate that an individual, blinded since birth or early childhood, is unlikely to gain functional vision even if sight is restored in adulthood. See

Gregory RL, Wallace J. Recovery from early blindness: A case study. In Gregory RL (ed.), *Concepts and Mechanisms of Perception.* London: Duckworth, 1974, 65–129.

Sacks OW. To see and not see. In *An Anthropologist on Mars.* New York: Alfred A. Knopf, 1995.

Von Senden M. *Sight and Space: The Perception of Space and Shape in the Congenitally Blind before and after Operation.* Glencoe, IL: Free Press, 1932.

However, Dr. Pawan Sinha, a neuroscientist at MIT, has studied patients in his native India who were blinded by cataracts or other conditions throughout early childhood. Some of the individuals in Sinha's study developed functional vision after surgery performed in late childhood or even adulthood.

Ostrovsky Y, Andalman A, Sinha P. Vision following extended congenital blindness. *Psychological Science* 17 (2006): 1009–14.

In addition, the following book recounts the struggle to regain vision by an individual who was blinded by a chemical accident at age three and regained sight as an adult.

Kurson R. *Crashing Through: A True Story of Risk, Adventure, and the Man Who Dared to See.* New York: Random House, 2007.

These reports indicate that significant visual plasticity does extend beyond a critical period in early life.

15 *Had I looked at papers and books written by a small subset of optometrists, I would have encountered clinicians who had developed procedures to rehabilitate people's vision, even the vision of individuals like me with a lifelong strabismus.*

Etting GL. Strabismus therapy in private practice: Cure rates after three months of therapy. *Journal of the American Optometric Association* 49 (1978): 1367–73.

Flax N, Duckman RH. Orthoptic treatment of strabismus. *Journal of the American Optometric Association* 49 (1978): 1353–61.

Ludlam WM. Orthoptic treatment of strabismus: A study of one hundred forty nine non-operated, unselected, concomitant strabismus patients completing orthoptic training at the Optometric Center of New York. *American Journal of Optometry and Archives of the American Academy of Optometry* 38 (1961): 369–88.

Ludlam WM, Kleinman BI. The long range results of orthoptic treatment of strabismus. *American Journal of Optometry and Archives of the American Academy of Optometry* 42 (1965): 647–84.

Press, LJ. *Applied Concepts in Vision Therapy.* St. Louis, MO: Mosby, 1997.

Chapter 2: Mixed-Up Beginnings

17 *My parents first noticed my misaligned eyes when I was only three months old.*

Clinical tests confirmed that my strabismus developed within the first year of life since, as an adult, I had latent nystagmus and an asymmetric optokinetic response with poor monocular optokinetic nystagmus in the nasotemporal direction.

18 *Since some children with crossed eyes straighten them spontaneously, the doctor suggested that my parents wait to see if I outgrew the condition.*

Cases in which the esotropia does not resolve usually involve infants with constant eye turns.

Fu VL, Stager DR, Birch EE. Progression of intermittent, small-angle, and variable esotropia in infancy. *Investigative Ophthalmology and Visual Science* 48 (2007): 661–64.

Pediatric Eye Disease Investigator Group. Spontaneous resolution of early-onset esotropia: Experience of the Congenital Esotropia Observational Study. *American Journal of Ophthalmology* 133 (2002): 109–18.

19   *My muscles were functioning fine, but the coordination of the two eyes was off.*

Lennerstrand G. Strabismus and eye muscle function. *Acta Ophthalmology Scandinavia* 85 (2007): 711–23.

Tychsen L, Richards M, Wong A, Foeller P, Burkhalter A, Narasimhan A, Demer J. Spectrum of infantile esotropia in primates: Behavior, brains, and orbits. *Journal of AAPOS* 12 (2008): 375–80.

The vast majority of infants who develop strabismus develop a non-paralytic form of esotropia. In a classic paper by FD Costenbader, 36 out of 1,152 infantile esotropes, or only 3.1 percent, had a paralytic form of strabismus.

Costenbader FD. Infantile esotropia. *Transactions of the American Ophthalmologial Society* 59 (1961): 397–429.

20   *A baby's sensory world is actually very different from an adult's.*

Daw NW. *Visual Development.* New York: Springer, 2006.

Slater A (ed.). *Perceptual Development: Visual, Auditory, and Speech Perception in Infancy.* East Sussex, UK: Psychology Press, 1998.

20   *But newborns have some innate perceptual skills—babies, at just nine minutes old, exhibit a preference for looking at a human face.*

Goren CC, Sarty M, Wu PY. Visual following and pattern discrimination of face-like stimuli by newborn infants. *Pediatrics* 56 (1975): 544–49.

20   *In addition, the eyes of a very young infant are not always straight.*

Nixon RB, Helveston EM, Miller K, Archer SM, Ellis FD. Incidence of strabismus in neonates. *American Journal of Ophthalmology* 100 (1985): 798–801.

Archer SM, Sondhi N, Helveston EM. Strabismus in infancy. *Ophthalmology* 96 (1989): 133–37.

Horwood A. Neonatal ocular misalignments reflect vergence development but rarely become esotropia. *British Journal of Ophthalmology* 87 (2003): 1146–50.

21   *To establish when the brain starts comparing the images from the two eyes, scientists at the Massachusetts Institute of Technology (MIT) placed Polaroid goggles on healthy babies whose parents had agreed to have them participate in experiments.*

Shimojo S, Bauer Jr J, O'Connell KM, Held R. Pre-stereoptic binocular vision in infants. *Vision Research* 26 (1986): 501–10.

22   *After four months, however, the brain determines whether the input comes from the right or left eye, and the babies begin to experience binocular rivalry.*

This period may correlate with the stage when afferents from the lateral geniculate nucleus have segregated into right-eye and left-eye ocular dominance columns in layer 4C of the visual cortex.

Shimojo S, Bauer Jr J, O'Connell KM, Held R. Pre-stereoptic binocular vision in infants. *Vision Research* 26 (1986): 501–10.

Daw NW. *Visual Development.* New York: Springer, 2006, 52–54.

23   *Scientists have surmised, therefore, that the ability to converge the eyes, to fuse two images together, and to appreciate stereoscopic depth may all develop at about the same time.*

Thorn F, Gwiazda J, Cruz AA, Bauer JA, Held R. The development of eye alignment, convergence, and sensory binocularity in young infants. *Investigative Ophthalmology and Visual Science* 35 (1994): 544–53.

Birch EE, Shimojo S, Held R. Preferential-looking assessment of fusion and stereopsis in infants aged 1–6 months. *Investigative Ophthalmology and Visual Science* 26 (1985): 366–70.

Daw NW. *Visual Development.* 2nd ed. New York: Springer, 2006.

23   *"Infantile esotropia" appears at about two to three months of age, while a second type of strabismus, "accommodative esotropia," usually develops later, at around two to three years.*

Archer SM, Sondhi N, Helveston EM. Strabismus in infancy. *Ophthalmology* 96 (1989): 133–37.

Nixon RB, Helveston EM, Miller K, Archer SM, Ellis FD. Incidence of strabismus in neonates. *American Journal of Ophthalmology* 100 (1985): 798–801.

Press LJ. *Applied Concepts in Vision Therapy.* St. Louis, MO: Mosby, 1997.

Von Noorden GK. *Binocular Vision and Ocular Motility.* 5th ed. St. Louis, MO: Mosby, 1996.

23   *There may be multiple causes of a poor ability to fuse and the development of crossed eyes.*

Daw NW. *Visual Development.* 2nd ed. New York: Springer, 2006.

Major A, Maples WC, Toomey S, DeRosier W, Gahn D. Variables associated with the incidence of infantile esotropia. *Optometry* 78 (2007): 534–41.

von Noorden GK. *Binocular Vision and Ocular Motility.* 5th ed. St. Louis, MO: Mosby, 1996, ch. 9.

24   *Infants learn about space through vision, touch, and their own movements.*

White BL, Castle P, Held R. Observations on the development of visually-directed reaching. *Child Development* 35 (1964): 349–64.

Bruner JS. *Processes of Cognitive Growth: Infancy.* Worcester, MA: Clark University Press, 1968.

Volume 37, number 3 (2006) of the journal *Optometry and Vision Development* is devoted almost entirely to infant vision.

Gesell A, Ilg FL, Bullis GE. *Vision: Its Development in Infant and Child.* Santa Clara, CA: Optometric Extension Program Foundation, 1998. (You can obtain books from the Optometric Extension Program via http://oep.excerpo.com.)

24   *In a classic study, scientists at MIT showed that accurate reaching in cats develops only when the kittens are able to watch their own limbs as they move them.*

Held R., Hein A. Movement-produced stimulation in the development of visually guided behavior. *Journal of Comparative and Physiological Psychology* 56 (1963): 872–76.

24   *Their developing visual skills reinforce their emerging motor skills and vice versa.*

Daw NW. *Visual Development.* 2nd ed. New York: Springer, 2006, ch. 3.

26   *"I see two images but only one is real."*

Observations similar to Sarah Merhar's are discussed in the following two papers:

McLaughlin SC. Visual perception in strabismus and amblyopia. *Psychological Monographs: General and Applied* 78 (1964): 1–23.

Brock FW. Space perception in its normal and abnormal aspects. *Optometric Weekly,* August 29 and September 5, 1946.

Sarah has subsequently undertaken optometric vision therapy with

Drs. Theresa Ruggiero and Cathy Stern, and her double vision has almost completely resolved.

28   *To make it easier to disregard one eye's input, I turned in the eye that was not doing the looking.*

When I undertook optometric vision therapy at age forty-eight, I could see this misalign-and-suppress mechanism at work in my own visual system. With therapy procedures, I learned to bring the images from both eyes into consciousness and could therefore discover where my two eyes were aiming. Let's say I looked at two dots, one red and one green, arranged side by side. While wearing red/green lenses, each color dot could be seen by only one eye. I might first look at the green dot with just the left eye open. Then, I would open both eyes. For a tiny fraction of a second, the red dot seen by the right eye would appear in the right place with respect to the green dot seen by the left eye. Then, very quickly, the red dot would move out of alignment. Presumably, throughout life, this unconscious action had moved the image from one eye out of alignment, making it easier for me to discount the image from the nonfixating eye.

28   *Dr. Fasanella could not give them an answer then, but recent research has shown that young infants, even with normal vision, can move each eye more effectively toward the nose than away from it.*

Atkinson J. Development of optokinetic nystagmus in the human infant and monkey infant. In R. D. Freeman (ed.), *Developmental Neurobiology of Vision.* New York: Plenum, 1979.

Naegele JR, Held R. The postnatal development of monocular optokinetic nystagmus in infants. *Vision Research* 22 (1982): 341–46.

Norcia AM. Abnormal motion processing and binocularity: Infantile esotropia as a model system for effects of early interruptions of binocularity. *Eye* 10 (1996): 259–65.

Teller DY, Succop A, Mar C. Infant eye movement asymmetries: Stationary counterphase gratings elicit temporal-to-nasal optokinetic nystagmus in two-month-old infants under monocular test conditions. *Vision Research* 33 (1993): 1859–64.

This mechanism may explain why crossed eyes are by far more common than walleye in very young babies. Results of studies with macaque monkeys support this idea. Macaques have a visual system similar to our own. In one study, prism goggles were put over the monkeys' eyes on their first day of life. The prism lenses shifted the visual field of one eye outward and that of the other eye upward. Now the visual fields of the two eyes were misaligned, preventing the possibility of binocular fusion and presumably causing double vision and visual confusion. Although the prisms shifted the visual field of one eye outward, the monkeys did not compensate by moving the eye outward. Instead, they turned in the eye behind the prism. If the prisms were left on the monkeys for the first twelve weeks of life, then they developed a constant esotropia. In this way, like children who cannot fuse, they may have developed a strategy to ignore or suppress the input from one eye. That strategy involved turning in the nonfixating eye.

Tychsen L. Causing and curing infantile esotropia in primates: The role of decorrelated binocular input. *Transactions of the American Ophthalmological Society* 105 (2007): 564–93.

29   *People who have been cross-eyed since early childhood see much less depth using motion parallax than people with normal binocular vision, and this, along with the lack of stereopsis, greatly compromises depth perception.*

Nawrot M, Frankl M, Joyce L. Concordant eye movement and motion parallax asymmetries in esotropia. *Vision Research* 48 (2008): 799–808.

Thompson AM, Nawrot M. Abnormal depth perception from motion parallax in amblyopic observers. *Vision Research* 39 (1999): 1407–13.

A poor sense of depth through motion parallax in infantile esotropes may result from poor pursuit movements of the eyes.

Naji JJ, Freeman TC. Perceiving depth order during pursuit eye movement. *Vision Research* 44 (2004): 3025–34.

Nawrot M, Joyce L. The pursuit theory of motion parallax. *Vision Research* 46 (2006): 4709–25.

Birch EE, Fawcett S, Stager D. Co-development of VEP motion response and binocular vision in normal infants and infantile esotropes. *Investigative Ophthalmology and Visual Science* 41 (2000): 1719–23.

Norcia AM. Abnormal motion processing and binocularity: Infantile esotropia as a model system for effects of early interruptions of binocularity. *Eye* 10 (1996): 259–65.

Tychsen L. Causing and curing infantile esotropia in primates: The role of decorrelated binocular input. *Transactions of the American Ophthalmological Society* 105 (2007): 564–93.

Tychsen L, Lisberger SG. Maldevelopment of visual motion processing in humans who had strabismus with onset in infancy. *Journal of Neuroscience* 6 (1986): 2495–508.

Tychsen L, Hurtig RR, Scott WE. Pursuit is impaired but the vestibulo-ocular reflex is normal in infantile strabismus. *Archives of Ophthalmology* 103 (1985): 536–39.

Valmaggia C, Proudlock F, Gottlob I. Optokinetic nystagmus in strabismus: Are asymmetries related to binocularity? *Investigative Ophthalmology and Visual Science* 44 (2003): 5142–50.

29   *As a result, many cross-eyed babies show delays in mastering tasks like grasping a toy or holding a bottle, and older children with the same problems may even show abnormalities in gait and posture.*

Birnbaum MH. Gross motor and postural characteristics of strabismic patients. *Journal of the American Optometric Association* 45 (1974): 686–96.

Drover JR, Stager DR, Morale SE, Leffler MN, Birch EE. Improvement in motor development following surgery for infantile esotropia. *Journal of AAPOS* 12 (2008): 136–40.

Slavik BA. Vestibular function in children with nonparalytic strabismus. *Occupational Therapy Journal of Research* 2 (1982): 220–33.

33   *This is true for many children with strabismus, particularly if they have surgery after the first year of life, the presumed critical period for the development of stereovision.*

In fact, the well-respected ophthalmologist and strabismic expert Stewart Duke-Elder wrote in 1976, "Insofar as the cure of squint [strabismus] is measured in terms of the restoration of binocular vision, *operation cannot by itself effect a cure*; it is merely a mechanical expedient to orientate the eyes."

Duke-Elder S, Wybar K. *System of Ophthalmology*, Vol. VI: *Ocular Motility and Strabismus.* St. Louis, MO: C. V. Mosby Co., 1973, 489.

Clinical studies indicate that no more than 20 percent of patients who undergo strabismic surgery after the age of two acquire stereopsis, while 20 to 80 percent of patients who receive surgery in the first year develop some binocular vision and stereopsis.

Birch EE, Felius J, Stager Sr DR, Weakley Jr DR, Bosworth RG. Pre-operative stability of infantile esotropia and post-operative outcome. *American Journal of Ophthalmology* 138 (2004): 1003–9.

Birch EE, Stager Sr DR. Long-term motor and sensory outcomes after early surgery for infantile esotropia. *Journal of AAPOS* 10 (2006): 409–13.

Birch EE, Stager DR, Everett ME. Random dot stereoacuity following surgical correction of infantile esotropia. *Journal of Pediatric*

Ophthalmology and Strabismus 32 (1995): 231–35.

Helveston EM, Neely DF, Stidham DB, Wallace DK, Plager DA, Spunger DT. Results of early alignment of congenital esotropia. *Ophthalmology* 106 (1999): 1716–26.

Hiles DA, Watson BA, Biglan AW. Characteristics of infantile esotropia following early bimedial rectus recession. *Archives of Ophthalmology* 98 (1980): 697–703.

Kushner BJ, Fisher M. Is alignment within 8 prism diopters of orthotropia a successful outcome for infantile esotropia surgery? *Archives of Ophthalmology* 114 (1996): 176–80.

Park MM. Stereopsis in congenital esotropia. *American Orthoptic Journal* 47 (1997): 99–102.

In the last twenty years, many investigators who study the effects of strabismic surgery on vision have concluded that the critical factor in achieving stereopsis may not be the *age* at the time of surgery but rather the *duration* of eye misalignment. This makes sense if we consider strabismic ways of seeing as adaptations to the disorder. The longer the eyes are misaligned, the longer the child has to cope with uncorrelated input from the two eyes and the longer is the period during which adaptations, such as suppression and anomalous correspondence, set in. If these adaptations provide the infant with a single view of the world and allow the baby to move with reasonable accuracy, then the child may not change his or her way of seeing even after surgery.

Mohindra I, Zwaan J, Held R, Brill S, Zwaan F. Development of acuity and stereopsis in infants with esotropia. *Ophthalmology* 92 (1985): 691–97.

Wright KW, Edelman PM, McVey JH, Terry AP, Lin M. High-grade stereo acuity after early surgery for congenital esotropia. *Archives of Ophthalmology* 112 (1994): 913–19.

Birch EE, Fawcett S, Stager DR. Why does early surgical alignment improve stereoacuity outcomes in infantile esotropia? *Journal of AAPOS* 4 (2000): 10–14.

Ing MR, Okino LM. Outcome study of stereopsis in relation to duration of misalignment in congenital esotropia. *Journal of AAPOS* 6 (2002): 3–8.

34   *Babies who can fuse images and develop stereopsis after surgery are more likely to keep their eyes aligned and require no further operations.*

Arthur BW, Smith JT, Scott WE. Long-term stability of alignment in the monofixation syndrome. *Journal of Pediatric Ophthalmology and Strabismus* 26 (1989): 224–31.

Chapter 3: School Crossings

38   *Yet, many school administrators and physicians have long questioned the connection between vision and learning.*

Helveston EM, Weber JC, Miller K, Robertson K, Hohberger G, Estes R, Ellis FD, Pick N, Helveston BY. Visual function and academic performance. *American Journal of Ophthalmology* 99 (1985): 346–55.

Learning disabilities, dyslexia, and vision. A subject review. *Pediatrics* 102 (1998): 1217–19. A joint policy statement of the American Academy of Pediatrics, the American Academy of Ophthalmology, and the American Association for Pediatric Ophthalmology and Strabismus. Available online at http://aappolicy.aappublications.org/cgi/content/full/pediatrics;102/5/1217

In contrast, the American Academy of Optometry and the American Optometric Association issued a joint statement indicating that vision-related reading problems do exist.

Vision, learning and dyslexia: A joint organizational policy statement of the American Academy of Optometry and the American Optometric Association. Available online at http://aaopt.org/userfiles/imagesPOSI TION_PAPERS_BV/img_3124863_Policy_Statement_On_Vision _Learning_And_Dyslexia.doc.

38   *Although the exact role of vision in learning is a subject of intense debate, many scientific studies support a connection between vision and reading.*

For review, see Kulp MT, Schmidt PP. Effect of oculomotor and other visual skills on reading performance: A literature review. *Optometry and Vision Science* 73 (1996): 283–92.

Cooper J. Summary of research on the efficacy of vision therapy for specific visual dysfunctions. *Journal of Behavioral Optometry* 9 (1998): 115–19. Also available online at http://visiontherapy.org/vision-therapy/vision-therapy-studies.html.

Griffin JR, Grisham JD. *Binocular Anomalies: Diagnosis and Vision Therapy.* New York: Butterworth-Heinemann, 2002.

Maples WC. Visual factors that significantly impact academic performance. *Optometry* 74 (2003): 35–49.

39   *For example, one paper published in 2007 examined the visual skills of 461 high school students who read at two or more levels below the established level for their grade.*

Grisham D, Powers M, Riles P. Visual skills of poor readers in high school. *Optometry* 78 (2007): 542–49.

39   *Additional papers have demonstrated a correlation between reading skill*
      level and the ability to see with stereopsis.

      Kulp MT, Schmidt PP. Visual predictors of reading performance in
      kindergarten and first grade children. *Optometry and Vision Science* 73
      (1996): 255–62.

      Kulp MT, Schmidt PP. A pilot study. Depth perception and near
      stereoacuity: Is it related to academic performance in young children?
      Binocular Vision and Strabismus Quarterly 17 (2002): 129–34.

      In addition, a lack of stereopsis in people with strabismus and am-
      blyopia is correlated with a poor ability to read letters on eye charts, that
      is, to interpret visually complicated patterns.

      McKee SP, Levi DM, Movshon JA. The pattern of visual deficits in
      amblyopia. *Journal of Vision* 3 (2003): 380–405.

39   *But recent experiments examining the movement of both eyes together*
      during reading have yielded some important surprises.

      Kirby JA, Webster LAD, Blythe SI, Liversedge SP. Binocular coordi-
      nation during reading and non-reading tasks. *Psychological Bulletin* 134
      (2008): 742–63.

      Liversedge SP, Rayner K, White SJ, Findlay JM, McSorley E. Binoc-
      ular coordination of the eyes during reading. *Current Biology* 16 (2006):
      1726–29.

40   *When I read with my right eye, the image of the word fell on the fovea*
      of my right eye and on the blind spot of my turned left eye.

      My reading adaptation is an example of the "blind spot mechanism"
      and has been observed in many other strabismics.

      Pratt-Johnson JA, Tillson G. *Management of Strabismus and Ambly-
      opia: A Practical Guide.* New York: Thieme Medical Pubs., 1994.

      Rutstein RP, Levi DM. The blind spot syndrome. *Journal of the Amer-
      ican Optometric Association* 50 (1979): 1387–90.

      For a conflicting opinion, see Olivier P, von Noorden GK. The blind
      spot syndrome: Does it exist? *Journal of Pediatric Ophthalmology and Stra-
      bismus* 18 (1981): 20–22.

      If I could move my eyes with enough precision to have the image of
      the letters fall on the blind spot of one eye, then why didn't I move my eyes
      instead in a way that caused the images to fall on the fovea of both eyes?
      The fovea is a much smaller target than the blind spot, so bifoveal fixation
      requires much more precise eye alignment than the strategy I used.

42   *After all these years of doctor's visits and school tests, Michelle finally*
      learned that Eric's difficulties resulted from a visual condition called con-

vergence insufficiency, a common but often undiagnosed cause of reading troubles.

Rouse MW, Borsting E, Hyman L, Hussein M, Cotter SA, Flynn M, Scheiman M, Gallaway M, De Land PN. Frequency of convergence insufficiency among fifth and sixth graders. The Convergence Insufficiency and Reading Study (CIRS) group. *Optometry and Vision Science* 76 (1999): 643–49.

Textbooks on optometric vision therapy that include descriptions of convergence insufficiency are listed below:

Griffin JR, Grisham JD. *Binocular Anomalies: Diagnosis and Vision Therapy.* New York: Butterworth-Heinemann, 2002.

Press, LJ. *Applied Concepts in Vision Therapy.* St. Louis, MO: 1997.

Scheiman M, Wick B. *Clinical Management of Binocular Vision: Heterophoric, Accommodative, and Eye Movement Disorders.* 2nd ed. New York: Lippincott Williams & Wilkins, 2002.

See also this reference for a *New York Times* account of the condition: Novak L. Not autistic or hyperactive: Just seeing double at times. *New York Times*, September 11, 2007. Available online at www.nytimes.com/2007/09/11/health/11visi.html.

Some individuals with convergence insufficiency manage to get through grade school without too many problems but "hit a wall" when greater demands are placed on their visual system in graduate school. Oliver Waldman and Garry Brown both suffered from convergence insufficiency, and both encountered significant problems when in law school or when taking the written medical boards, respectively. It was at that point that they learned about, and engaged in, optometric vision therapy, and only then were they able to realize their career goals.

42 *According to a recent National Eye Institute study of 221 children, the most effective treatment for Eric's condition is a combination of office-based and home-based vision therapy.*

Convergence Insufficiency Treatment Trial Study Group. Randomized clinical trial of treatments for symptomatic convergence insufficiency in children. *Archives of Ophthalmology* 126 (2008): 1336–49.

See the National Eye Institute press release on the above report at http://covd.org/Portals/0/NEIPressRelease.pdf.

Scheiman M, Mitchell GL, Cotter S, Cooper J, Kulp M, Rouse M, Borsting E, London R, Wensveen J. Convergence Insufficiency Treatment Trial Study Group. A randomized clinical trial of treatments for convergence insufficiency in children. *Archives of Ophthalmology* 123 (2005): 14–24.

Chapter 4: Knowing Where to Look

47   *In his thoughtful and moving memoir,* Touching the Rock, *John Hull recounts what it is like to be blind.*
       Hull JM. *Touching the Rock: An Experience of Blindness.* New York: Pantheon Books, 1990.

47   *If we didn't have legs and arms for climbing trees and fingers for manipulating objects, we would never have needed such a complicated visual system or brain.*
       Many optometrists, scientists, and philosophers have recognized the important connection between vision and movement.
       Birnbaum MH. Behavioral optometry: A historical perspective. *Journal of the American Optometric Association* 65 (1994): 255–64.
       Churchland PS, Ramachandran VS, Sejnowski TJ. A critique of pure vision. In Koch C, Davis JL (eds.), *Large Scale Neuronal Theories of the Brain.* Cambridge, MA: MIT Press, 1994. Available online at http://philosophy.ucsd.edu/faculty/pschurchland/papers/kochdavis94critique ofpurevision.pdf.
       Daw NW. *Visual Development.* 2nd ed. New York: Springer, 2006, 31.
       Gibson JJ. *The Ecological Approach to Visual Perception.* Hillsdale, NJ: Lawrence Erlbaum Associates, 1986.
       Harmon DB. *Notes on a Dynamic Theory of Vision.* Santa Ana, CA: Optometric Extension Program Foundation, 1958.
       Noe E. *Action in Perception.* Cambridge, MA: MIT Press, 2004.

48   *You gaze directly at the curve a second or two before rounding the bend, then turn your head in the direction of your gaze and masterfully steer the car around the curve.*
       Land MF, Lee DN. Where we look when we steer. *Nature* 369 (1994): 742–44.

50   *Your sensed visual direction—the direction in which you are looking— is not the direction in which either eye is pointing but one that seems to emanate from the center of your forehead.*
       Brock FW. A comparison between strabismic seeing and normal binocular vision. *Journal of the American Optometric Association* 31 (1959): 299–304.

50   *For most people, images that fall on corresponding retinal points appear in the same subjective visual direction.*

This observation was first described in the second century AD by Ptolemy in Books II and III of his *Optics*. See Howard IP, Rogers BJ. *Seeing in Depth*. Ontario: I. Porteus, 2002, vol. 1, ch. 2.

52   *While this turn of events may seem surprising, similar cases have been documented in which the amblyopic eye regains vision following severe impairment of the "good" eye.*
El Mallah MK, Chakravarthy U, Hart PM. Amblyopia: Is visual loss permanent? *British Journal of Ophthalmology* 84 (2000): 952–56.
Tierney DW. Vision recovery in amblyopia after contralateral subretinal hemorrhage. *Journal of the American Optometric Association* 60 (1989): 281–83.
Vereecken EP, Brabant P. Prognosis for vision in amblyopia after the loss of the good eye. *Archives of Ophthalmology* 102 (1984): 220–24.
Wilson ME. Adult amblyopia reversed by contralateral cataract formation. *Journal of Pediatric Ophthalmology and Strabismus* 29 (1992): 100–102.

54   *Bruce's unusual way of seeing is called anomalous correspondence, an adaptation that often takes months or years of childhood strabismus to develop.*
Brock FW. Investigation into anomalous correct projection in cases of concomitant squints. *American Journal of Optometry* 16 (1939): 39–77.
Brock FW. Anomalous projection in squint. Its cause and effect. New methods of correction. Report of cases. *American Journal of Optometry* 16 (1939): 201–21.
Brock FW. Space perception in its normal and abnormal aspects. *Optometric Weekly* 37 (1946): 1193–96, 1202, 1235–38.
Steinman SB, Steinman BA, Garzia RP. *Foundations of Binocular Vision: A Clinical Perspective*. New York: McGraw-Hill Cos., 2000.
Von Noorden GK. *Binocular Vision and Ocular Motility*, 5th ed. New York: Mosby, 1996, ch. 13.

55   *In his book* The Organism, *Goldstein stresses that a patient's symptoms are often a response to, or coping mechanism for dealing with, his or her disorder.*
Goldstein K. *The Organism: A Holistic Approach to Biology Derived from Pathological Data in Man*. New York: Zone Books, 1995.

55   *With Goldstein's research in mind, Brock noted that a strabismic "speaks a different language."*
Brock FW. Conditioning the squinter to normal visual habits. *Optometric Weekly* 32 (1941): 793–801, 819–24.

57   *Like Rachel, my friend Tracy Gray suffered from a twisted posture asso-
      ciated with her vision.*
      Darell Boyd Harmon did extensive studies on the connections be-
      tween vision and posture among 160,000 school children.
      Harmon DB. *Notes on a Dynamic Theory of Vision.* Santa Ana, CA:
      Optometric Extension Program Foundation, 1958 (obtained via
      http://oep.excerpo.com).

62   *Instead, it is built upon orthoptic procedures developed in the late 1800s
      by the French ophthalmologist Louis Emile Javal.*
      Duke-Elder S, Wybar K. *System of Ophthalmology.* Vol. VI: *Ocular
      Motility and Strabismus.* St. Louis, MO: C. V. Mosby Co., 1973.
      Griffin JR, Grisham JD. *Binocular Anomalies: Diagnosis and Vision
      Therapy.* New York: Butterworth-Heinemann, 2002, ch. 9.
      Javal E. *Manuel du strabisme.* Paris: Masson, 1986.
      Press LJ. *Applied Concepts in Vision Therapy.* New York: Mosby, 1997,
      ch. 1.
      Revell MJ. *Strabismus: A History of Orthoptic Techniques.* London: Bar-
      rie & Jenkins, 1971, 15–23.

62   *As mentioned in chapter 1, laboratory experiments indicate that eye mis-
      alignment during a "critical period" in early life disrupts the development
      of binocular neurons.*
      Daw NW. Critical periods and strabismus: What questions remain?
      Optometry and Vision Science 74 (1997): 690–94.
      Crawford MLG, Harwerth RS, Smith EL, von Noorden GK. Keep-
      ing an eye on the brain: The role of visual experience in monkeys and
      children. *Journal of General Psychology* 120 (1993): 7–19.
      Daw NW. Critical periods in the visual system. In Hopkins B, John-
      son SP (eds.), *Neurobiology of Infant Vision.* Westport, CT: Praeger Pub.,
      2003.
      Daw NW. *Visual Development.* New York: Springer, 2006, ch. 9.
      Hubel DH, Wiesel TN. Binocular interaction in striate cortex of kit-
      tens reared with artificial squint. *Journal of Neurophysiology* 28 (1965):
      1041–59.
      Hubel DH, Wiesel TN. *Brain and Visual Perception: The Story of a 25-
      Year Collaboration.* Oxford: Oxford University Press, 2005.

62   *As a result, ophthalmologists, in the late 1900s, began to operate on stra-
      bismic infants within the first year of life.*

Tychsen L. Can ophthalmologists repair the brain in infantile esotropia? Early surgery, stereopsis, monofixation syndrome, and the legacy of Marshall Parks. *Journal of AAPOS* 9 (2005): 510–21.

62   **Surgery on such young children has been partially successful in allowing for the development of stereovision.**

Birch EE, Fawcett S, Stager DR. Why does early surgical alignment improve stereoacuity outcomes in infantile esotropia? *Journal of AAPOS* 4 (2000): 10–14.

Birch EE, Felius J, Stager Sr DR, Weakley Jr DR, Bosworth RG. Preoperative stability of infantile esotropia and post-operative outcome. *American Journal of Ophthalmology* 138 (2004): 1003–9.

Birch EE, Stager Sr DR. Long-term motor and sensory outcomes after early surgery for infantile esotropia. *Journal of AAPOS* 10 (2006): 409–13.

Ing MR, Okino LM. Outcome study of stereopsis in relation to duration of misalignment in congenital esotropia. *Journal of AAPOS* 6 (2002): 3–8.

Park MM. Stereopsis in congenital esotropia. *American Orthoptic Journal* 47 (1997): 99–102.

Wright KW, Edelman PM, McVey JH, Terry AP, Lin M. High-grade stereo acuity after early surgery for congenital esotropia. *Archives of Ophthalmology* 112 (1994): 913–19.

63   **The babies are simply too young to participate in vision therapy procedures.**

However, a caretaker (parent, babysitter, or therapist) can perform some therapy procedures with a cross-eyed infant to encourage abduction (moving the eyes outward), promote awareness of the visual periphery, and reduce cross-fixation (the tendency to use the right eye to see the left side of the visual field and vice versa.)

Gingham fabrics can be used on crib bumpers, chair coverings, and wall paper. The repeating patterns on these fabrics present the same stimulus pattern to both eyes, even eyes that are misaligned, and may help promote fusion. See Press LJ. *Applied Concepts in Vision Therapy.* St. Louis, MO: Mosby, 1997, 99–100.

63   **In a study to determine the effectiveness of this training, 149 patients (none of whom had undergone any surgery) received treatment sessions twice a week for twelve weeks.**

Ludlam WM. Orthoptic treatment of strabismus: A study of one hundred forty-nine non-operated, unselected, concomitant strabismus patients completing orthoptic training at the Optometric Center of New York. *American Journal of Optometry and Archives of the American Academy of Optometry* 38 (1961): 369–88.

Ludlam WM, Kleinman BI. The long range results of orthoptic treatment of strabismus. *American Journal of Optometry and Archives of the American Academy of Optometry* 42 (1965): 647–84.

63   *Several other investigators confirmed these important and groundbreaking studies.*

Etting GL. Strabismus therapy in private practice: Cure rates after three months of therapy. *Journal of the American Optometric Association* 49 (1978): 1367–73.

Flax N, Duckman RH. Orthoptic treatment of strabismus. *Journal of the American Optometric Association* 49 (1978): 1353–61.

63   *Optometrists have always been at the forefront of lens development, designing, for example, the first contact lenses as well as low-vision devices (tools for reading and distance viewing used by people with severe vision loss).*

Schaeffer J. Contact lens pioneers. *Review of Optometry* 144 (2007). Available online at http://www.revoptom.com/index.asp?ArticleType= SiteSpec&page=contactlens/index.htm.

64   *In 1971, a subgroup of optometrists particularly interested in vision therapy founded the College of Optometrists in Vision Development (COVD) in order to standardize therapy protocols and develop a rigorous test for board certification in optometric vision therapy.*

Press LJ. *Applied Concepts on Vision Therapy.* New York: Mosby, 1997, ch. 1.

Information on COVD can be found at http://covd.org.

In addition, the American Academy of Optometry has instituted a diplomate program of the Section on Binocular Vision and Perception for optometrists who specialize in binocular vision disorders.

65   *Dr. Ruggiero went on to explain that my eyes, though cosmetically straight, were still both horizontally and vertically misaligned.*

Following are results from my optometric tests performed on November 20, 2001.

Wirt circles	No measurable stereoacuity
Random dot E test	Fail
Keystone skills	Alternating fixation, no superimposition or fusion
Worth 4 dot	Left-eye suppression at intermediate and far distance (twenty feet); double vision (diplopia) at near (sixteen inches)
Red lens test	Initially left-eye suppression and then diplopia
Maddox rod	Right hypertropia of 5.5 prism diopters at distance, 4 prism diopters at intermediate, and 3 prism diopters at near
Cover test	Constant left-eye esotropia at distance, intermediate, and near. 8 prism diopters of left esotropia at far; 25 prism diopters of left esotropia at near with right hypertropia at all viewing distances
Fusion	Binocular fusion not possible at any viewing distance or direction of gaze; simultaneous awareness led to diplopia
Motility	No restriction of gaze; latent nystagmus
Refraction	Right eye: −1.00–0.75 x 55 Left eye: −1.25–0.50 x 65
Acuity	Right eye, w/o glasses: 20/50; w/ glasses: 20/20 Left eye, w/o glasses: 20/60; w/ glasses: 20/20 Both eyes w/glasses: 20/20

Chapter 5: Fixing My Gaze

70   *In fact, the act of planning the movement, which is largely unconscious, and the movement itself may sensitize our eyes, ears, and fingers.*
      Gibson JJ. Observations on active touch. *Psychological Review* 69 (1962): 477–91.
      Rosenbaum DA. *Human Motor Control.* San Diego, CA: Academic Press, 1991, 23.

70   *In a series of extraordinary experiments, Paul Bach-y-Rita and his colleagues demonstrated the importance of self-directed movement in perceiving the world.*

Bach-y-Rita P. Tactile sensory substitution studies. *Annals of the New York Academy of Sciences* 1013 (2004): 83–91.

71   *In fact, this important skill involves a large number of brain regions.*
For books on gaze holding and eye movements, consider
Berthoz A. *The Brain's Sense of Movement.* Cambridge, MA: Harvard University Press, 2000.
Leigh RJ, Zee DS. *The Neurology of Eye Movements.* Philadelphia: F. A. Davis Co., 1991.

71   *So, when you look steadily at an object, you must continually refresh the image by moving your eyes subtly, for instance, by slowly sweeping them across the object and making small, quick jerks.*
Daw NW. *Visual Development.* 2nd ed. New York: Springer, 2006, 31.
Martinez-Conde S, Macknik SL, Hubel DH. Microsaccadic eye movements and firing of single cells in the striate cortex of monkeys. *Nature Neuroscience* 3 (2000): 251–8.
Martinez-Conde S, Macknik SL, Hubel DH. The role of fixational eye movements in visual perception. *Nature Reviews. Neuroscience* 5 (2004): 229–40.

73   *This abnormal movement, called latent nystagmus, is often seen in people who have been cross-eyed since infancy.*
Brodsky MC. Visuo-vestibular eye movements: Infantile strabismus in three dimensions. *Archives of Ophthalmology* 123 (2005): 837–42.
Richards M, Wong A, Foeller P, Bradley D, Tychsen L. Duration of binocular decorrelation predicts the severity of latent (fusion maldevelopment) nystagmus in strabismic macaque monkeys. *Investigative Ophthalmology and Visual Science* 49 (2008): 1872–78.
Tychsen L. Infantile esotropia: Current neurophysiologic concepts. In Rosenbaum AL, Santiago AP (eds.), *Clinical Strabismus Management.* Philadelphia: W. B. Saunders, 1999, 117–38.

74   *Smooth pursuit movements of the eyes are often abnormal in people who have been cross-eyed since infancy, and this deficit may result from poor development of stereovision.*
Birch EE, Fawcett S, Stager D. Co-development of VEP motion response and binocular vision in normal infants and infantile esotropes. *Investigative Ophthalmology and Visual Science* 41 (2000): 1719–23.
Norcia AM. Abnormal motion processing and binocularity: Infantile esotropia as a model system for effects of early interruptions of binocularity. *Eye* 10 (1996): 259–65.

Tychsen L, Lisberger SG. Maldevelopment of visual motion processing in humans who had strabismus with onset in infancy. *Journal of Neuroscience* 6 (1986): 2495–508.

Tychsen L, Hurtig RR, Scott WE. Pursuit is impaired but the vestibulo-ocular reflex is normal in infantile strabismus. *Archives of Ophthalmology* 103 (1985): 536–39.

Valmaggia C, Proudlock F, Gottlob I. Optokinetic nystagmus in strabismus: Are asymmetries related to binocularity? *Investigative Ophthalmology and Visual Science* 44 (2003): 5142–50.

75 *When some astronauts first return from space, they complain that the world appears to move when they turn their heads or walk, an effect that could cause serious problems if they have to perform emergency procedures right after landing.*

Reschke MF, Bloomberg JJ, Harm DL, Paloski WH. Space flight and neurovestibular adaptation. *Journal of Clinical Pharmacology* 34 (1994): 609–17.

Reschke MF, Bloomberg JJ, Harm DL, Paloski WH, Layne C, McDonald V. Posture, locomotion, spatial orientation, and motion sickness as a function of space flight. *Brain Research Reviews* 28 (1998): 102–17.

77 *The astronauts who adapt most easily to spaceflight are the "head lockers," those who move their eyes, head, and body in unison when they are first afloat in space.*

Amblard B, Assaiante C, Vaugoyeau M, Baroni G, Ferrigno G, Pedotti A. Voluntary head stabilisation in space during oscillatory trunk movements in the frontal plane performed before, during and after a prolonged period of weightlessness. *Experimental Brain Research* 137 (2001): 170–79.

78 *Optometrist A. M. Skeffington is often considered the father of vision therapy.*

Francke AW. Our optometric heritage. *Visions* (COVD newsletter) 38 (2008): 4.

78 *Observations like this one led optometrists Amiel Francke and Robert Kraskin to design balance boards in the mid-1900s to use in vision therapy.*

Francke AW. Our optometric heritage. *Visions* (COVD newsletter) 38 (2008): 4.

79 *When asked how he did this, he said that when you play on a basketball court long enough, "you develop a sense of where you are."*

McPhee J. *A Sense of Where You Are: A Profile of William Warren Bradley*. New York: Farra, Straus and Giroux, 1978, 22.

80   *In* Living in a World Transformed, *Hubert Dolezal examines what it is like to see without peripheral vision by wearing a set of tubes over his eyes for one week.*
Dolezal H. *Living in a World Transformed.* New York: Academic Press, 1982, 57–79.

84   *I find the looking-soft technique useful in all sorts of places.*
My improved vision helps me appreciate music in a number of other ways. I sing in a chorus, and as I learned to make better use of my peripheral vision, I was able to follow the musical score and the conductor's baton at the same time. I have always been a pretty sloppy pianist, but as my vision improved, I could read the sheet music with my central vision, while using my peripheral vision to move my hands accurately around the keyboard. I began to play Chopin's *Grand Waltz*, a piece that I had thought was beyond me.

Although I did not receive vision therapy until age forty-eight, learning to play the piano as a child may have helped me with my future vision training. My father sometimes insisted that I play a piece perfectly three times in a row before I could get up from the piano bench. This led to some tearful afternoons. While this kind of discipline may have been excessive, it did teach me how to practice. I learned to break the difficult passages in the music down into smaller parts, work on each, and then put them back together again into a musical whole. I realized that I could get better with practice, a concept that we all know but often do not embrace or follow. Since playing the piano requires using the two hands differently, I learned to pay attention to both sides of my body at the same time. What's more, playing the piano developed my sense of rhythm. Learning all of these skills contributed to my later success with vision training in a way that all the book learning in school never could have.

85   *Certainly, more fingers are moved and more work is done while opening the whole hand, but it takes more neuronal input to lift only one finger.*
Schieber MH. How might the motor cortex individuate movement? *Trends in Neurosciences* 13 (1990): 440–45.

86   *Suppression is strongest under natural, daytime viewing conditions, under conditions when a strabismic most needs a single view of the world.*
Jampolsky A. Characteristics of suppression in strabismus. *AMA Archives of Ophthalmology* 54 (1955): 683–96.

McLaughlin SC. Visual perception in strabismus and amblyopia. *Psychological Monographs: General and Applied* 78 (1964): 1–23.

Press LJ. *Applied Concepts in Vision Therapy*. New York: Mosby, 1997, 211.

88 *If I turned on the input from both eyes, I asked Dr. Ruggiero, wouldn't I see double?*

In fact, many patients, such as Sarah Merhar, Pat Duffy, mentioned in chapter 8, and Margaret Lundin, mentioned below, were able to reduce or eliminate their double vision using optometric vision therapy.

In 1973, when Margaret Lundin was twenty-one years old, she had a benign but large tumor removed from her left frontal sinus. When she woke up from the operation, she had double vision. Glasses with prisms and eye muscle surgery reduced, but did not eliminate, her double vision. Margaret recently consulted Dr. Ruggiero, who began vision therapy with her, and for the first time in over thirty years, she learned to coordinate her eyes in order to see a single, stereoscopic view of the world.

Chapter 6: The Space Between

89 *It must be repeated here that, before stereopsis is actually experienced by the patient . . .*

Brock FW. Anomalous projection in squint. Its cause and effect. New methods of correction. Report of cases. *American Journal of Optometry* 16 (1939): 201–21.

89 *The disparity between these two images was too great for me to automatically make corrective convergence or divergence eye movements.*

Studies on strabismics indicate that they do not make normal vergence movements.

Burian HM. Fusional movements in permanent strabismus. *Archives of Ophthalmology* 26 (1941): 626–52.

Kenyon RV, Ciuffreda KJ, Stark L. Dynamic vergence eye movements in strabismus and amblyopia: Symmetric vergence. *Investigative Ophthalmology and Visual Science* 19 (1980): 60–74.

Kenyon RV, Ciuffreda KJ, Stark L. Dynamic vergence eye movements in strabismus and amblyopia: Asymmetric vergence. *British Journal of Ophthalmology* 65 (1981): 167–76.

Schor CM, Ciuffreda KJ. *Vergence Eye Movements: Basic and Clinical Aspects*. Boston: Butterworth Pubs., 1983.

90   *Not content to test merely what his patients could not see, he performed many thoughtful experiments to determine what and how his patients did see.*

Brock FW. Investigation into anomalous correct projection in cases of concomitant squints. *American Journal of Optometry* 16 (1939): 39–77.

Brock FW. Anomalous projection in squint. Its cause and effect. New methods of correction. Report of cases. *American Journal of Optometry* 16 (1939): 201–21.

Brock FW. Binocular vision in strabismus. *Optometric Weekly* 35 (1945): 1417–18; 36 (1945): 67–68, 179–80, 291–93, 401–3, 575–77, 687–89, 773–75, 1099–100, 1131–33, 1253–56; 37 (1946): 71–74, 231–36.

Brock FW. Space perception in its normal and abnormal aspects. *Optometric Weekly* 37 (1946): 1193–96, 1202, 1235–38.

Greenwald I. *Effective Strabismus Therapy.* Santa Ana, CA: Optometric Extension Program Foundation, 1979, available online at http://oep.excerpo.com.

Greenwald I. *Strabismus: Brock's Influence on New Therapies.* Santa Ana, CA: Optometric Extension Program Foundation, 1982/3, vols. 1 and 2, available at http://oep.excerpo.com.

90   *In so doing, he taught us that you can learn a lot from a simple piece of string.*

Brock FW. The string as an aid to visual training. *Optometric Extension Program: Visual Training at Work,* series 4, no. 9 (1955): 29–33.

92   *This area is called Panum's fusional area, and I could also fuse the images of those parts of the string that fell within this area.*

Mitchell DE. A review of the concept of "Panum's fusional areas." *American Journal of Optometry and Archives of American Academy of Optometry* 43 (1966): 387–401.

Ogle KN. *Researches in Binocular Vision.* Philadelphia: W. B. Saunders Co., 1950.

93   *Movement alone enhances our perception.*

Gibson JJ. Observations on active touch. *Psychological Review* 69 (1962): 477–91.

Rosenbaum DA. *Human Motor Control.* San Diego, CA: Academic Press, 1991, 23.

94   *Most importantly, he realized that stable, clear binocular vision and stereopsis could be achieved only if the strabismic actively positioned his or her eyes, or made what Brock called a "fusion effort."*
        Brock FW. The fusion range in stereoscopic vision: Part 6: The perception of depth. *Optometric Weekly* 33 (1942): 777–79.
        Brock realized that a strabismic was unlikely to achieve fusion and see in stereoscopic depth as long as his eyes were in a strabismic posture. In classical orthoptic procedures, the patient may look into a stereoscope or amblyoscope whose arms have been adjusted for his strabismic angle. While the patient looks through the stereoscope, the same image is cast on both foveas. Since the patient's eyes are still in a strabismic posture, however, he will have a difficult time overcoming his strabismic adaptations in order to fuse the images. Brock's techniques differed from classical orthoptics because he emphasized that the patient must move his eyes out of the strabismic posture into better alignment, and only then was fusion and stereopsis possible. The Brock string was one tool I used to learn how to align my eyes.
        Like me, Jennifer Clark transformed her view of the world by discovering and changing the way she moved her eyes. Jennifer is from Gourock, Scotland, has a PhD in biology, and for several years worked in a scientific laboratory. Then, at age twenty-eight, she got a job in bioinformatics, which led to hours of computer work. At that point, Jennifer told me, her "near vision just shut down." Soon she could barely read at all. She had only one area of clear vision off to the side. Jen went on a search for a doctor who could help her and finally found optometrist Carolyn Hurst of St. Neots, England. Caroline diagnosed Jennifer's problem as a severe case of double vision with convergence insufficiency and accommodative dysfunction.
        When we look at a near object, we do more than turn in our eyes. The lenses of our eyes also change shape, or, in technical terms, accommodate, so that the object that we are fixating appears in sharp focus. The coupling of these two processes, aiming and focusing the eyes, develops in the first year of life, is fine-tuned throughout childhood, and works automatically for most people. Jennifer was unable to converge her eyes to fuse the image from each eye and see singly, and so everything she saw was doubled and blurred. As she had never seen a completely clear single image, Jennifer hadn't learned to use the accommodation (focusing ability) of her eyes to keep the image clear at any distance.
        One day, four months into her vision therapy, Jennifer had a revelation. Caroline Hurst had explained that she could use the Brock string to determine where her eyes were aiming since this was the point on the

string where the two string images crossed. Jennifer discovered for herself that when she looked down the string, the point most sharply in focus was not the point at which the string images crossed. She now had a way to determine the plane in which her eyes converged and the plane in which her eyes were focused. She realized that instead of trying to look at the place where the two string images crossed, she should be looking at the part of the string that was most in focus. Very soon after introducing this modification, she was able to get the focused part of the string to move up to the place where the string images crossed. Jennifer was now converging at the same place on the string where she was focusing. It was shortly after she mastered this task that she began to see in 3D.

My story and first experience with stereopsis has been described in Oliver Sacks's article, "Stereo Sue." *The New Yorker*, June 19, 2006, 64–73.

My story was also described by Robert Krulwich in a *Morning Edition* program on National Public Radio. See "Going Binocular: Susan's First Snowfall," June 26, 2006, at www.npr.org/templates/story/story.php?storyId=5507789.

96 *During this period, I reread* The Man Who Mistook His Wife for a Hat.
Sacks O. *The Man Who Mistook His Wife for a Hat.* New York: Summit Book, 1985, 56–62.

99 *How strong or weak an individual synaptic connection becomes depends upon when and with whom it is active.*
This basic idea was first formulated by Donald Hebb in 1949.
Hebb DO. *The Organization of Behavior: A Neuropsychological Theory.* New York: John Wiley & Sons, 1949.
An excellent review of neuronal correlates of learning can be found in Kandel ER. *In Search of Memory: The Emergence of a New Science of Mind.* New York: W. W. Norton & Co., 2006.

99 *By a process called long-term potentiation, the previously ineffective connection between the left-eye pathway and the postsynaptic neuron can be strengthened, perhaps to the point where stimulation of the left-eye neuron alone can get the postsynaptic cell to fire.*
In the laboratory, synaptic changes due to long-term potentiation have been well studied in the adult mammalian visual cortex. See Artola A, Singer W. Long-term depression of excitatory synaptic transmission and its relationship to long-term potentiation. *Trends in Neurosciences* 16 (1993): 480–87.

Artola A, Brocher S, Singer W. Different voltage-dependent thresholds for inducing long-term depression and long-term potentiation in slices of rat visual cortex. *Nature* 347 (1990): 69–72.

Kirkwood A, Bear MF. Hebbian synapses in visual cortex. *Journal of Neuroscience* 14 (1994): 1634–45.

Kirkwood A, Bear MF. Homosynaptic long-term depression in the visual cortex. *Journal of Neuroscience* 14 (1994): 3404–12.

101 *Now a binocular neuron, even a very weakly binocular neuron, received correlated input from the two eyes.*

In young cats, covering one eye results in the loss of binocular neurons in the visual cortex and in reduced visual acuity of the covered eye. However, brief periods of normal binocular vision, which provides correlated binocular input to visual cortical cells, promote recovery of normal visual acuity in the deprived eye.

Mitchell DE. A special role for binocular visual input during development and as a component of occlusion therapy for treatment of amblyopia. *Restorative Neurology and Neuroscience* 26 (2008): 425–34.

Kind PC, Mitchell DE, Ahmed B, Blakemore C, Bonhoeffer T, Sengpiel F. Correlated binocular activity guides recovery from monocular deprivation. *Nature* 416 (2002): 430–33.

Mitchell DR, Kind PC, Sengpiel F, Murphy K. Short periods of concordant binocular vision prevent the development of deprivation amblyopoia. *European Journal of Neuroscience* 23 (2006): 2458–66.

Schwarzkopf DS, Vorbyov V, Mitchell DE, Sengpiel F. Brief daily binocular vision prevents monocular deprivation effects in visual cortex. *European Journal of Neuroscience* 25 (2007): 270–80.

101 *For example, one eye may have blocked or inhibited connections from the other eye onto a visual neuron, and this inhibitory effect may have been reduced when I was able to aim both eyes accurately at the same place in space.*

An additional mechanism for the loss of binocularity in strabismics may be the formation of inhibitory connections between the right- and left-eye pathways. These connections may prevent a neuron from receiving input from both eyes. Only the uninhibited pathway causes the so-called monocular neuron to fire. Indeed, in laboratory experiments, application of drugs that block inhibition reveals that 30 to 50 percent of the monocular neurons in the brains of strabismic cats actually receive connections from both eyes. When I gained stereovision, inhibitory connections between the pathways from the two eyes may have been reduced.

Burchfiel JL, Duffy FH. Role of intracortical inhibition in deprivation amblyopia: Reversal by microiontophoretic bicuculline. *Brain Research* 16 (1981): 479–84.

Chino YM, Smith III EL, Kazuyuki Y, Cheng H, Hamamoto J. Binocular interactions in striate cortical neurons of cats reared with discordant visual input. *Journal of Neuroscience* 14 (1994): 5050–67.

Fagiolini M, Hensch TK. Inhibitory threshold for critical-period activation in primary visual cortex. *Nature* 404 (2000): 183–86.

Hensch TK. Critical period plasticity in local cortical circuits. *Nature Reviews. Neuroscience* 6 (2005): 877–88.

Mower GD, Christen WG, Burchfiel JL, Duffy FH. Microiontophoretic bicuculline restores binocular responses to visual cortical neurons in strabismic cats. *Brain Research* 309 (1984): 168–72.

Sengpiel F, Blakemore C. The neural basis of suppression and amblyopia in strabismus. *Eye* 10 (1996): 250–58.

Sillito AM, Kemp JA, Blakemore C. The role of GABAergic inhibition in the cortical effects of monocular deprivation. *Nature* 28 (1981): 318–20.

Sillito AM, Kemp JA, Patel H. Inhibitory interactions contributing to the ocular dominance of monocularly dominated cells in the normal cat striate cortex. *Experimental Brain Research* 41 (1980): 1–10.

Smith EL, Chino YM, Ni J, Cheng H, Crawford ML, Harwerth RS. Residual binocular interactions in the striate cortex of monkeys reared with abnormal binocular vision. *Journal of Neurophysiology* 78 (1997): 1353–62.

Furthermore, drugs or conditions that alter synaptic inhibition may also modulate plasticity in the visual cortex.

Maya Ventencourt JF, Sale A, Viegi A, Baroncelli L, De Pasquale R, O'Leary OF, Castren E, Maffei L. The antidepressant fluoxetine restores plasticity in the adult visual cortex. *Science* 320 (2008): 385–88.

Sale A, Maya Vetencourt JF, Medini P, Cenni MC, Baroncelli L, De Pasquale R, Maffei L. Environmental enrichment in adulthood promotes amblyopia recovery through a reduction of intracortical inhibition. *Nature Neuroscience* 10 (2007): 679–81.

102 In their book Phantoms in the Brain, *V. S. Ramachandran and S. Blakeslee define the term* quale *(plural:* qualia*) as "the raw feel of sensations such as the subjective quality of 'pain' or 'red' or 'gnocchi with truffles.'"*

Ramachandran VS, Blakeslee S. *Phantoms in the Brain: Probing the Mysteries of the Human Mind.* New York: Quill William Morrow, 1998.

103   *Frederick Brock was a strabismic.*
Dr. Brock was an intermittent exotrope. Brock FW. *Lecture Notes on Strabismus.* Meadville, PA: Keystone View Co.

Chapter 7: When Two Eyes See As One

107   *In fact, by discovering and exploiting these cues, the artists of the past were in many ways vision scientists.*
Fascinating books on vision and art include the following:
Gregory RL, Gombrich EH. *Illusion in Nature and Art.* New York: Charles Scribner's Sons, 1973.
Livingstone M. *Vision and Art: The Biology of Seeing.* New York: Harry N. Abrams, 2002.
Zeki S. *Inner Vision.* New York: Oxford University Press, 1999.

112   *He came to visit and, a year later, wrote an article, "Stereo Sue," for* The New Yorker *magazine.*
Sack O. Stereo Sue. *The New Yorker,* June 19, 2006, 64–73.

112   *A week after the article's publication, I was interviewed on National Public Radio.*
"Going Binocular: Susan's First Snowfall." June 26, 2006, *Morning Edition* on National Public Radio. Available online at www.npr.org/templates/story/story.php?storyId=5507789.

115   *Fusing large images, called peripheral fusion, is an important step in gaining stereopsis because it helps align the eyes and trigger accurate convergence and divergence movements.*
Burian HM. Fusional movements: Role of peripheral retinal stimuli. *Archives of Ophthalmology* 21 (1939): 486–91.
Brock FW. Pitfalls in orthoptic training of squints. *Optometric Weekly* 32 (1941): 1185–89.

118   *I experienced another curious effect with the rope circle vectogram that optometrists call the small in, large out (SILO) phenomenon.*
Press LJ. *Applied Concepts in Vision Therapy.* St. Louis, MO: Mosby, 1997, 232–34.
Scheiman M, Wick B. *Clinical Management of Binocular Vision: Heterophoric, Accommodative, and Eye Movement Disorders.* 2nd ed. Philadelphia: Lippincott Williams & Wilkins, 2002, 130–33.

121   *I was quite happy about "getting" some random dot stereograms because many scientists believe that seeing images in these stereograms, images*

*that cannot be seen with monocular cues alone, is the ultimate proof that
a person has stereopsis.*
Bela Julesz pioneered the use of random dot stereograms in perceptual studies.
    Julesz B. *Foundations of Cyclopean Perception*. Chicago: University of
Chicago Press, 1971.
Random dot stereograms are often used as a clinical test for stereopsis, but I question whether these stereograms provide the best measure
of the emergence of stereopsis in individuals with amblyopia and strabismus. Not only must the patient use retinal disparity cues to see random dot stereograms, but he must give these cues top priority in his
interpretation of the figure. If the patient has been stereoblind since early
childhood, he has depended upon monocular cues to interpret distance
and depth. These monocular cues may continue to dominate even after
he gains stereopsis, and monocular cues will tell him that the stereogram
is a flat drawing. In his interpretation of real objects in the real three-
dimensional world, monocular and retinal disparity cues usually provide
correlated, not conflicting, information and will combine to give a view
seen in compelling depth.
    In addition, many strabismics and amblyopes suffer from the crowd-
ing phenomenon. They can identify a letter on an eye chart more easily
if the letter is seen in isolation rather than being flanked by other letters.
    Irvine RS. Amblyopia ex anopsia. Observations on retinal inhibition,
scotoma, projection, light difference discrimination and visual acuity.
Transactions of the American Ophthalmological Society 46 (1948): 527–75.
    Von Noorden GK. *Binocular Vision and Ocular Motility*. New York:
Mosby, 1996, 225–28.
    Levi DM, Song S, Pelli D. Amblyopic reading is crowded. *Journal of
Vision* 7 (2007): article 21, 1–17.
The crowding phenomenon can make it difficult to see individual
dots or clusters of dots in random dot stereograms. With this in mind,
Westheimer and McKee came up with a stereopsis test that, like random
dot stereograms, lacks monocular cues but can be seen by someone who
has stereopsis but experiences crowding.
    Westheimer G, McKee SP. Stereogram design for testing local stere-
opsis. *Investigative Ophthalmology and Visual Science* 19 (1980): 802–
809.
Finally, it is possible to see random dot stereograms even with poor
stereovision. For example, Tara Fitzpatrick is an exotrope who had surgery that cosmetically aligned her eyes. However, she has had no vision
therapy. When she fixates with one eye, the other eye turns out, although

the turn is not noticeable to the casual observer. She can pull her eyes into position to see a random dot stereogram. However, posturing her eyes for stereovision takes a great deal of effort, so she does not normally do this. Instead, she uses one eye for distance viewing and the other eye for near. Although she can see a random dot stereogram, she does not see like a person with normal stereovision.

123 *This new sense of immersion in space is completely captivating and enchanting.*

I was fascinated to read on the Internet a piece by music critic Nick Coleman, who lost hearing in one ear. Music now seemed flat to him. It lost its emotionality, its spaciousness; it no longer surrounded and inhabited him. He even quoted from "Stereo Sue," Oliver Sacks' account of my story, in which I described my joy at being "inside" a snowfall. The sense of immersion that I gained when I learned to see with two eyes is what he lost when he could no longer hear through two ears.

Coleman N. Life in mono. *Guardian*, February 19, 2008. Available online at www.guardian.co.uk/lifeandstyle/2008/feb/19/healthandwellbeing.classicalmusicandopera.

123 *Indeed, Rachel Cooper, who had a form of amblyopia (lazy eye), notes that before gaining stereopsis, "It felt like I was here and everything I was looking at was over there."*

See online at http://children-special-needs.org/lazy_eye/lazy_eye.html.

125 *Although the estimate of the object's exact location or depth is not precise, you have an impression of its "nearerness" or "furtherness."*

This sensation is called qualitative stereopsis.

Ogle KN. Disparity limits of stereopsis. *AMA Archives of Ophthalmology* 48 (1952): 50–60.

Westheimer G, Tanzman IJ. Qualitative depth localization with diplopic images. *Journal of the Optical Society of America* 46 (1956): 116–17.

Brock FW. Visual training—Part III. *Optometric Weekly* 46–50 (1955–59).

Brock FW. A comparison between strabismic seeing and normal binocular vision. *Journal of the American Optometric Association* 31 (1959): 299–304.

125 *It was this newfound sense of stereo depth for objects all around me that gave me the powerful feeling of being enveloped by the world.*

Margaret Corbit, a fifty-nine-year-old artist with amblyopia, also described to me how her sense of space changed when she consulted a developmental optometrist and recently gained stereovision. Like me, she used to see the world in a few discrete planes. There might be a flat plane of objects seen in detail close to her, another plane at mid-distance, and then a vague flat background. She drew and painted that way too. Margaret also struggled to draw in perspective, but without a good sense of depth and distance, she was not particularly bothered by this problem. In fact, she hesitated to go for vision therapy because she was not sure that gaining stereopsis would be good for her art. She worried that seeing in stereo would compel her to render things in proper depth and perspective and that this would take the fun out of drawing. Now she can turn her stereopsis on and off and is captivated by her new stereovision and sense of space. So far, her experience is enriching her playful treatment of these concepts in her art.

126 *I learned that the same neurons and circuits that give us stereopsis may also provide us with our sensation of depth through motion parallax.*

Bradley CD, Qian N, Anderson RA. Integration of motion and stereopsis in middle temporal cortical area of macaques. *Nature* 373 (1995): 609–11.

Pack CC, Born RT, Livingstone MS. Two-dimensional substructure of stereo and motion interactions in macaque visual cortex. *Neuron* 37 (2003): 525–35.

Roy JP, Komatsu H, Wurtz RH. Disparity sensitivity of neurons in monkey extrastriate area MST. *Journal of Neuroscience* 12 (1992): 2478–92.

Upadhyay UD, Page WK, Duffy CJ. MST responses to pursuit across optic flow with motion parallax. *Journal of Neurophysiology* 84 (2000): 818–26.

126 *What's more, experiments by neurobiologist Mark Nawrot and his colleagues at North Dakota State University have identified the signals coming into the brain that provide us with our sense of depth through motion parallax.*

Nawrot M, Joyce L. The pursuit theory of motion parallax. *Vision Research* 46 (2006): 4709–25.

Also see Naji JJ, Freeman TC. Perceiving depth order during pursuit eye movement. *Vision Research* 44 (2004): 3025–34.

Psychophysical experiments on humans also indicate that depth perception from stereopsis and motion parallax are not independent processes.

Bradshaw MF, Rogers BJ. The interaction of binocular disparity and motion parallax in the computation of depth. *Vision Research* 36 (1996): 3457–68.

126   The same researchers have also reported that individuals with crossed eyes and amblyopia have a poor sense of depth through motion parallax.
Nawrot M, Frankl M, Joyce L. Concordant eye movement and motion parallax asymmetries in esotropia. *Vision Research* 48 (2008): 799–808. Thompson AM, Nawrot M. Abnormal depth perception from motion parallax in amblyopic observers. *Vision Research* 39 (1999): 1407–13.

127   So, I went back to the library to read about structure from motion and was not surprised to learn that the capacities to see depth through stereopsis and to determine structure from motion are linked.
Richards W, Lieberman HR. Correlation between stereo ability and the recovery of structure-from-motion. *American Journal of Optometry and Physiological Optics* 62 (1985): 111–18.

130   With my new outlook, all of my senses were awakened.
These experiences were beautifully described by Rebecca Penneys, an accomplished concert pianist and professor of piano whose eyes began to cross at age two. Although she had three surgeries, she did not gain stereovision until she engaged in optometric vision therapy with optometrist Ray Gottlieb in her forties. She writes,

> Thinking back on it all again makes me want to reiterate that having full depth perception affected my whole personality and being, and was one of the most positive and incredible experiences of my life. It changed my hearing, my perceptual abilities in general, made me happier. I was able to learn faster and understood everything more completely from that moment on. It was an enormous "a-ha." The *stress* of seeing was replaced by the *joy* of sight.

131   Most surprising to me was that the change in my vision affected the way that I thought.
A fascinating discussion of the connection between vision and thought can be found in the following book:
Arnheim R. *Visual Thinking.* Berkeley: University of California Press, 1969.

Chapter 8: Nature and Nurture

133   Although clinical tests in Dr. Ruggiero's office confirmed that I now saw in stereo, I was not totally convinced.
Following are the records contrasting the results of my clinical tests on November 20, 2001 and January 22, 2008.

	November 20, 2001	January 22, 2008
Wirt circles	No measurable stereoacuity	Stereoacuity of 70 arcseconds
Random dot E test	Fail	Pass
Keystone skills	Alternating fixation, no superimposition or fusion	Binocular vision with superimposition and fusion
Worth 4 dot	Left-eye suppression at intermediate and far distance (20 feet). Double vision (diplopia) at near (16 inches)	See all four dots to fifteen feet.
Cover test	Constant left-eye esotropia at distance, intermediate, and near. 8 Prism diopters of left esotropia at far; 25 prism diopters of left esotropia at near with right hypertropia at all viewing distances	Orthophoric at all viewing distances; small right hyperphoria
Fusion	Binocular fusion not possible at any viewing distance; simultaneous awareness led to diplopia	Fusion at all viewing distances and directions of gaze
Refraction	Right eye: −1.00–0.75 x 55 Left eye: −1.25–0.50 x 65 Near vision: +1.25	Right eye: −1.25–0.50 x 62 Left eye: −1.50–0.75 x 72 Near vision: +2.50
Visual acuity	Right eye: w/o glasses: 20/50; w/glasses: 20/20 Left eye: w/o glasses: 20/60; w/glasses: 20/20 Both eyes w/glasses: 20/20	Right eye: w/o glasses: 20/30; w/ glasses: 20/20 Left eye: w/o glasses: 20/40; w/glasses: 20/20 Both eyes w/glasses: 20/16

133   Some doctors argued that I must have had Duane's syndrome.
    For information on Duane's syndrome, see Von Noorden GK. *Binocular Vision and Ocular Motility.* 5th ed. New York: Mosby, 1996, 430–37.

134 *In the late 1800s, two great scientists, Ewald Hering and Hermann Von Helmholtz, hotly debated the nature-versus-nurture question with regard to vision.*
Howard IP, Rogers BJ. *Seeing in Depth.* Ontario: I. Porteus, 2002, 41–46.
Turner SR. Vision studies in Germany: Helmholtz versus Hering. *Osiris* 8 (1993): 80–103.

135 *At that time, treatment for crossed eyes was heavily influenced by Claud Worth, an ophthalmologist and author of* Squint: Its Causes, Pathology and Treatment.
Worth C. *Squint: Its Causes, Pathology, and Treatment.* Philadelphia: P. Blakiston's Son & Co., 1903. Available online at http://books .google.com/books?id=I0cSAAAAYAAJ&dq=worth,+squint&printsec=fr ontcover&source=bl&ots=fOhT4UTb4n&sig=parvBDlDOo_c8g Eck-lSfudZnT8&hl=en&sa=X&oi=book_result&resnum=1&ct=result# PPP1,M1.
Van Noorden GK. The development of the art and science of strabismology outside North America: Part II. *Journal of AAPOS* 5 (2001): 134–38.

135 *But in 1939, thirty-six years after the first edition of* Squint *was published, Chavasse challenged Worth's theory by postulating that the ability to fuse depended upon a series of reflexes that developed during childhood.*
Chavasse FB. *Worth's Squint on the Binocular Reflexes and the Treatment of Strabismus.* 7th ed. London: Bailliere, Tindall and Cox, 1939.

135 *For example, in a paper published in 1951, developmental biologist and Nobel laureate Roger Sperry demonstrated that neuronal circuitry between the eye and brain in fish and amphibians was "hardwired."*
Sperry RW. Developmental patterning of neural circuits. *Chicago Medical School Quarterly* 12 (1951): 66–73.
Sperry RW. The eye and the brain. *Scientific American* 194 (1956): 48–52.
Both of these articles are available online at www.rogersperry.info.

136 *One such behavior, known as imprinting, was first noted by the naturalist Douglas Alexander Spalding in 1873 and described in detail in the 1930s by the great animal behaviorist Konrad Lorenz.*
Lorenz K. The companion in the bird's world. *Auk* 54 (1937): 245–73. Available online at http://elibrary.unm.edu/sora/Auk/v054n03/ p0245-p0273.pdf.

Spalding DA. Instinct, with original observations on young animals. *Macmillan's Magazine* 27 (1873): 282–93; reprinted in *Animal Behavior* 2 (1954): 2–11.
A delightful book by Lorenz on animal behavior is
Lorenz KZ. *King Solomon's Ring*. New York: Thomas Y. Crowell Co., 1952.

136 **These were among the observations and ideas under discussion by scientists at the time that David Hubel and Torsten Wiesel began their studies of the mammalian visual system.**
Hubel DH, Wiesel TN. *Brain and Visual Perception: The Story of a 25-Year Collaboration*. Oxford: Oxford University Press, 2005.

137 **By observing the responses of individual neurons to patterns of light, Hubel and Wiesel uncovered some of the underlying neural mechanisms for vision.**
Hubel DH. *Eye, Brain, and Vision*. New York: Scientific American Library, 1995. Also available online at http://hubel.med.harvard.edu/bcontex.htm.

138 **In the primary visual cortex, Hubel and Wiesel discovered binocular neurons.**
Other investigators have extended the monocular/binocular classification of neurons by looking at the response of neurons to simultaneous input from both eyes. Some neurons, for example, may appear monocular if the stimulus is presented to only one eye at a time. Thus, a given neuron may respond, for example, if only the left eye is stimulated and, according to Hubel and Wiesel's original scheme, would be classified as monocular. However, the same neuron may show a stronger or weaker response if the stimulus is presented to both eyes simultaneously rather than just to one eye. Thus, input from both eyes actually affects the neuron's response. See, for example, Ohzawa I, Freemam RD. The binocular organization of simple cells in the cat's visual cortex. *Journal of Neuroscience* 56 (1986): 243–59.
Chino YM, Smith III EL, Yoshida K, Cheng H, Harmamoto J. Binocular interactions in striate cortical neurons of cats reared with discordant visual inputs. *Journal of Neuroscience* 14 (1994): 5050–67.

138 **In 1965, they published a pivotal paper on strabismus.**
Hubel DH, Wiesel TN. Binocular interaction in striate cortex of kittens reared with artificial squint. *Journal of Neurophysiology* 28 (1965): 1041–59.

138 *Particularly striking to the scientists was that these effects were found only in young animals.*
Yinon U. Age dependence of the effect of squint on cells in kittens' visual cortex. *Experimental Brain Research* 26 (1976): 151–57.

Wiesel TN. Nobel lecture: The postnatal development of the visual cortex and the influence of environment (1981). In Hubel DH, Wiesel TN. *Brain and Visual Perception: The Story of a 25-Year Collaboration.* Oxford: Oxford University Press, 2005.

139 *They found that about half of the children developed some stereovision if their eyes were aligned and stayed aligned during the first year of life.*
Costenbacher FD. Infantile esotropia. *Transactions of the American Ophthalmological Society* 59 (1961): 397–429.

Taylor DM. Is congenital esotropia functionally curable? *Transactions of the American Ophthalmological Society* 70 (1972): 529–76.

Ing MR. Early surgical alignment for congenital esotropia. *Transactions of the American Ophthalmological Society* 79 (1981): 625–63.

See also the following later studies: Birch EE, Fawcett S, Stager DR. Why does early surgical alignment improve stereoacuity outcomes in infantile esotropia? *Journal of AAPOS* 4 (2000): 10–14.

Birch EE, Felius J, Stager Sr DR, Weakley Jr DR, Bosworth RG. Preoperative stability of infantile esotropia and post-operative outcome. *American Journal of Ophthalmology* 138 (2004): 1003–9.

Birch EE, Stager Sr DR. Long-term motor and sensory outcomes after early surgery for infantile esotropia. *Journal of AAPOS* 10 (2006): 409–13.

Birch EE, Stager DR, Everett ME. Random dot stereoacuity following surgical correction of infantile esotropia. *Journal of Pediatric Ophthalmology and Strabismus* 32 (1995): 231–35.

Helveston EM, Neely DF, Stidham DB, Wallace DK, Plager DA, Spunger DT. Results of early alignment of congenital esotropia. *Ophthalmology* 106 (1999): 1716–26.

Hiles DA, Watson BA, Biglan AW. Characteristics of infantile esotropia following early bimedial rectus recession. *Archives of Ophthalmology* 98 (1980): 697–703.

Ing MR, Okino LM. Outcome study of stereopsis in relation to duration of misalignment in congenital esotropia. *Journal of AAPOS* 6 (2002): 3–8.

Kushner BJ, Fisher M. Is alignment within 8 prism diopters of orthotropia a successful outcome for infantile esotropia surgery? *Archives of Ophthalmology* 114 (1996): 176–80.

Park MM. Stereopsis in congenital esotropia. *American Orthoptic Journal* 47 (1997): 99–102.

140 *"A missing aspect of this work is knowledge of the time course of the strabismus animals, cats or monkeys, and in the monkeys the possibilities of recovery."*

Hubel DH, Wiesel TN. *Brain and Visual Perception*. New York: Oxford University Press, 2005, 590.

As Hubel had mentioned to me, animal studies have not clearly established a critical period for strabismus. To truly delineate a critical period, you must demonstrate that the effects of strabismus on cortical wiring cannot be reversed after a certain age. As in my case, surgical repair may not be enough to promote a change in vision. To replicate my visual experience in a monkey, you would have to teach the animal to perform vision therapy, a hard and tedious task. But two years ago, a research group at the University of Houston School of Optometry performed this sort of experiment. They made infant monkeys optically strabismic with prisms. The prisms shifted the visual field of the two eyes in such a way that the animals could not fuse images. When the prisms were removed from the animals at age fourteen weeks, they were stereoblind. At the age of two years, that is, as young adults, the stereoblind monkeys were given stereovision training. Following training, the strabismic monkeys improved their stereoacuity. What's more, the number of binocular neurons in their visual cortex increased. This experiment indicates that the proper kind of training or therapy can change cortical wiring even in an adult individual who has been stereoblind since early infancy.

Nakatsuka C, Zhang B, Watanabe I, Zheng J, Bi H, Ganz L, Smith EL, Harwerth RS, Chino YM. Effects of perceptual learning on local stereopsis and neuronal responses of V1 and V2 in prism-reared monkeys. *Journal of Neurophysiology* 97 (2007): 2612–26.

140 *For example, binocular neurons sensitive to retinal disparity are found in one-week-old macaque monkeys, the earliest age at which recordings of nerve impulses can be made.*

Chino YM, Smith III EL, Hatta S, Cheng H. Postnatal development of binocular disparity sensitivity in neurons of the primate visual cortex. *Journal of Neuroscience* 17 (1997): 296–307.

In addition, binocular, disparity-sensitive neurons have been found in newborn lambs that have not been exposed to light.

Clarke PGH, Ramachandran VS, Whitteridge FRS. The development of the binocular depth cells in the secondary visual cortex of the lamb. *Proceedings of the Royal Society of London B* 204 (1979): 455–65.

Ramachandran VS, Clarke PD, Whitteridge D. Cells selective to binocular disparity in the cortex of newborn lambs. *Nature* 268 (1977): 333–35.

Finally, binocular vision may be dependent upon the formation of ocular dominance columns in the visual cortex, and these columns complete formation shortly after birth.

Horton JC, Hocking DR. An adult-like pattern of ocular dominance columns in striate cortex of newborn monkeys prior to visual experience. *Journal of Neuroscience* 16 (1996): 1791–807.

LeVay S, Wiesel TN, Hubel DH. The development of ocular dominance columns in normal and visually deprived monkeys. *Journal of Comparative Neurology* 191 (1980): 1–51.

140   *Although babies are probably born with binocular neurons, they do not demonstrate a capacity for stereovision until about four months of age.*
Birch EE, Shimojo S, Held R. Preferential-looking assessment of fusion and stereopsis in infants aged 1–6 months. *Investigative Ophthalmology and Visual Science* 26 (1985): 366–70.

Fox R, Aslin RN, Shea SL, Dumais ST. Stereopsis in human infants. *Science* 207 (1980): 323–24.

Petrig B, Julesz B, Kropfl W, Baumgartner G, Anliker M. Development of stereopsis and cortical binocularity in human infants: Electrophysiological evidence. *Science* 213 (1981): 1402–5.

Thorn F, Gwiazda J, Cruz AA, Bauer JA, Held R. The development of eye alignment, convergence, and sensory binocularity in young infants. *Investigative Ophthalmology and Visual Science* 35 (1994): 544–53.

141   *So, I was intrigued to read studies published in the mid-1980s by Eileen Birch and David Stager of the Retina Foundation of the Southwest.*
Birch EE, Stager DR. Monocular acuity and stereopsis in infantile esotropia. *Investigative Ophthalmology and Visual Science* 26 (1985): 1624–30.

Stager DR, Birch EE. Preferential-looking acuity and stereopsis in infantile esotropia. *Journal of Pediatric Ophthalmology and Strabismus* 23 (1986): 160–65.

Also see Bechtoldt HP, Hutz CS. Stereopsis in young infants and stereopsis in an infant with congenital esotropia. *Journal of Pediatric Ophthalmology and Strabismus* 16 (1979): 49–54.

142   *Their experiments demonstrate that strabismic individuals who flunk the standard stereopsis tests used in the eye doctor's office may be able to see in 3D when they look at large or moving targets located in their peripheral visual fields.*

Kitaoji H, Toyama K. Preservation of position and motion stereopsis in strabismic subjects. *Investigative Ophthalmology and Visual Science* 28 (1987): 1260–67.

O'Shea RP, McDonald AA, Cumming A, Pearl D, Sanderson G, Molteno AC. Interocular transfer of the movement aftereffect in central and peripheral vision of people with strabismus. *Investigative Ophthalmology and Visual Science* 35 (1994): 313–17.

Sireteanu R. Binocular vision in strabismic humans with alternating fixation. *Vision Research* 22 (1982): 889–96.

Sireteanu R, Fronius M, Singer W. Binocular interaction in the peripheral visual field of humans with strabismic and anisometropic amblyopia. *Vision Research* 21 (1981): 1065–74.

143   *Indeed, Frederick Brock noted, "Nearly all strabismics have occasional moments when they maintain binocular vision. The only reason this is not generally known is that most of us [eye doctors] have never taken the trouble to discover the fact."*

Brock FW. *Lecture Notes on Strabismus*. Meadville, PA: Keystone View Co., n.d.

143   *One way that Dr. Brock and other optometrists promoted this latent stereovision in their patients was to project large stereo targets onto the wall.*

Brock FW. Pitfalls in orthoptic training of squints. *Optometric Weekly* 32 (1941): 1185–89.

143   *The vast majority of these experiments have involved recordings of neurons that respond to only the central 5° of the visual field.*

One study did report recordings made from noncentral neurons in a strabismic animal, and these neurons remained binocular.

Sengpiel F, Blakemore C. The neural basis of suppression and amblyopia in strabismus. *Eye* 10 (1996): 250–58.

144   *Optometrist Paul Harris calls this spectrum the binocular continuum.*

Harris P. The binocular continuum. *Journal of Behavioral Optometry* 13 (2002): 99–103.

144 *How far a person can move along this spectrum depends in large part upon how he has adapted to his visual disorder.*

In general, strabismics who have a very large eye turn or who demonstrate an entrenched anomalous correspondence will have a more difficult time gaining normal binocular vision.

Etting GL. Strabismus therapy in private practice: Cure rates after three months of therapy. *Journal of the American Optometric Association* 49 (1978): 1367–73.

Flax N, Duckman RH. Orthoptic treatment of strabismus. *Journal of the American Optometric Association* 49 (1978): 1353–61.

Ludlam WM. Orthoptic treatment of strabismus: A study of one hundred forty-nine non-operated, unselected, concomitant strabismus patients completing orthoptic training at the Optometric Center of New York. *American Journal of Optometry and Archives of the American Academy of Optometry* 38 (1961): 369–88.

Ludlam WM, Kleinman BI. The long range results of orthoptic treatment of strabismus. *American Journal of Optometry and Archives of the American Academy of Optometry* 42 (1965): 647–84.

Is it possible, however, that some people are simply born with deficient binocular wiring, and that is why they develop strabismus and never see in 3D? Do people with a family history of strabismus inherit a deficient number of binocular neurons and a poor binocular system? Surprisingly, patients with a family history of strabismus achieve as much or more success with optometric vision therapy than those with no family history. So, even these patients probably begin life with binocular neurons and have the potential to develop stable binocular vision with stereopsis.

Flom MC. Issues in the clinical management of binocular anomalies. In Rosenbloom AA, Morgan MW (eds.), *Principles and Practice of Pediatric Optometry*. Philadelphia: JB Lippincot, 1990.

Press, LJ. *Applied Concepts in Vision Therapy*. St. Louis, MO: Mosby, 1997, 96.

146 *The best way to test this was to simulate congenital cataracts in an animal and then make recordings from visual neurons.*

Wiesel TN, Hubel DH. Single-cell responses in striate cortex of kittens deprived of vision in one eye. *Journal of Neurophysiology* 26 (1963): 1003–17.

Hubel DH, Wiesel TN, LeVay S. Plasticity of ocular dominance columns in monkey striate cortex. *Philosophical Transactions of the Royal Society of London B* 278 (1977): 377–409.

LeVay S, Wiesel TN, Hubel DH. The development of ocular dominance columns in normal and visually deprived monkeys. *Journal of Comparative Neurology* 191 (1980): 1–51.

Hubel and Wiesel's experiments indicate that a child born with a cataract should have surgery to remove the cataract within the first months of life, a practice now commonly followed.

Birch EE, Stager DR. Prevalence of good visual acuity following surgery for congenital unilateral cataract. *Archives of Ophthalmology* 106 (1988): 40–43.

146 *If, while the animals were still very young, the situation was reversed so that the covered eye was opened and the open eye was occluded, then a change occurred in the cortical neurons.*

Blakemore C, Van Sluyters RC. Reversal of the physiological effects of monocular deprivation in kittens: Further evidence of a sensitive period. *Journal of Physiology* 237 (1974): 195–216.

LeVay S, Wiesel TN, Hubel DH. The development of ocular dominance columns in normal and visually deprived monkeys. *Journal of Comparative Neurology* 191 (1980): 1–51.

147 *They mistakenly assume that the "critical period" for the development of amblyopia is the same as the "critical period" for its rehabilitation.*

There are different critical periods for the development of various neuronal properties, such as direction selectivity and binocularity, and for the development of different perceptual properties, such as acuity and stereopsis. In addition, the critical period for the development of a particular property may be different from the critical period for the disruption of that property or for recovery from disruption.

Daw NW. *Visual Development*. 2nd ed. New York: Springer, 2006, ch. 9.

147 *Amblyopia can be caused by strabismus when only one eye does the looking.*

Daw NW. *Visual Development*. New York: Springer, 2006, ch. 8.

Ciuffreda KJ, Levi DM, Selenow A. *Amblyopia: Basic and Clinical Aspects*. Boston: Butterworth-Heinemann, 1991.

148 *So, unlike the animals in the monocular deprivation experiments, children with strabismic amblyopia have not been deprived of all vision in one eye since the first days of life.*

In Hubel and Wiesel's studies, kittens that had been made strabismic did not develop amblyopia but alternated fixation between the two eyes instead. Thus, Hubel and Wiesel were not able to develop a cat model for strabismic amblyopia. In a later study, they were able to produce one monkey with strabismic amblyopia.

Hubel DH, Wiesel TN. Binocular interaction in striate cortex of kittens reared with artificial squint. *Journal of Neurophysiology* 28 (1965): 1041–59.

Wiesel TN. The postnatal development of the visual cortex and the influence of environment. *Nature* 299 (1982): 583–91. (This Nobel lecture was delivered on December 8, 1981, and is also reprinted in Hubel DH, Wiesel TN. *Brain and Visual Perception: The Story of a 25-Year Collaboration.* Oxford: Oxford University Press, 2005, 686–704; see also 591.)

148   *Under these conditions, a condition called anisometropic amblyopia may develop in the more farsighted eye.*

Daw NW. *Visual Development.* New York: Springer, 2006, ch. 8.

Ciuffreda KJ, Levi DM, Selenow A. *Amblyopia: Basic and Clinical Aspects.* Boston: Butterworth-Heinemann, 1991.

148   *Not surprisingly, these investigators found that the changes in circuitry in the visual cortex were not as severe as in the cortex of the monocularly deprived animals.*

Kiorpes L, Boothe RG, Hendrickson AE, Movshon JA, Eggers HM, Gizzi MS. Effects of early unilateral blur on the macaques' visual system. I. Behavioral observations. *Journal of Neuroscience* 7 (1987): 1318–26.

Hendrickson AE, Movshon JA, Eggers HM, Gizzi MS, Boothe RG, Kiorpes L. Effects of early unilateral blur on the macaque's visual system. II. Anatomical observations. *Journal of Neuroscience* 7 (1987): 1327–39.

Movshon JA, Eggers HM, Gizzi MS, Hendrickson AE, Kiorpes L, Boothe RG. Effects of early unilateral blur on the macaques' visual system. III. Physiological observations. *Journal of Neuroscience* 7 (1987): 1340–51.

Kiorpes L, Kiper D, O'Keefe L, Cavanaugh J, Movshon J. Neuronal correlates of amblyopia in the visual cortex of macaque monkeys with experimental strabismus and anisometropia. *Journal of Neuroscience* 18 (1998): 6411–24.

See also Eggers HM, Blakemore C. Physiological basis of anisometropic amblyopia. *Science* 201 (1978): 264–67.

Horton JC, Stryker MP. Amblyopia induced by anisometropia without shrinkage of ocular dominance columns in human striate cortex.

224   Notes for Chapter 8: Nature and Nurture

Proceedings of the National Academy of Sciences 90 (1993): 5494–98.

149   *One drawback of occlusion therapies is that the open eye may improve*
      its acuity at the expense of the occluded, or "good," eye.

      Birch EE, Stager DR, Berry P, Everett ME. Prospective assessment of
      acuity and stereopsis in amblyopic infantile esotropes following early sur-
      gery. *Investigative Ophthalmology and Visual Science* 31 (1990): 758–65.

149   *Indeed, in his address upon winning the Nobel Prize for his research on*
      vision, Torsten Wiesel expressed these concerns.

      Wiesel TN. The postnatal development of the visual cortex and the
      influence of environment. *Nature* 299 (1982): 583–91. (This Nobel lec-
      ture was delivered on December 8, 1981, and is also reprinted in Hubel
      DH, Wiesel TN. *Brain and Visual Perception: The Story of a 25-Year Col-
      laboration.* Oxford: Oxford University Press, 2005, 686–704.)

      In addition, a study of 427 amblyopic adults suggests that a lack of
      normal binocular function and stereovision is correlated with a poor abil-
      ity to identify letters on eye charts (optotype, or Snellen, acuity), a task
      that requires interpretation of complex visual patterns.

      McKee SP, Levi DM, Movshon JA. The pattern of visual deficits in
      amblyopia. *Journal of Vision* 3 (2003): 380–405.

      Furthermore, in formerly monocular-deprived kittens, just brief
      periods of normal binocular vision are sufficient to promote the recov-
      ery of normal visual acuity in the deprived eye. These periods of normal
      vision provide correlated binocular input onto visual cortical neurons.

      Mitchell DE. A special role for binocular visual input during devel-
      opment and as a component of occlusion therapy for treatment of am-
      blyopia. *Restorative Neurology and Neuroscience* 26 (2008): 425–34.

      Kind PC, Mitchell DE, Ahmed B, Blakemore C, Bonhoeffer T,
      Sengpiel F. Correlated binocular activity guides recovery from monocu-
      lar deprivation. *Nature* 416 (2002): 430–33.

      Mitchell DE, Kind PC, Sengpiel F, Murphy K. Short periods of con-
      cordant binocular vision prevent the development of deprivation am-
      blyopia. *European Journal of Neuroscience* 23 (2006): 2458–66.

      Schwarzkopf DS, Vorbyov V, Mitchell DE, Sengpiel F. Brief daily
      binocular vision prevents monocular deprivation effects in visual cortex.
      European Journal of Neuroscience 25 (2007): 270–80.

149   *Scientific studies have revealed that the amblyopic eye not only has re-
      duced acuity but provides a distorted sense of space, tracks objects less ac-
      curately, and is the poorer eye for directing hand movements.*

Bedell HE, Flom MC. Monocular spatial distortion in strabismic amblyopia. *Investigative Ophthalmology and Visual Science* 20 (1981): 263–68.

Bedell HE, Flom MC. Normal and abnormal space perception. *American Journal of Optometry and Physiological Optics* 60 (1983): 426–35.

Daw NW. *Visual Development.* New York: Springer, 2006, ch. 8.

Hess RF, Campbell FW, Greenhalgh T. On the nature of the neural abnormality in human amblyopia: Neural aberrations and neural sensitivity loss. *Pflugers Archives* 377 (1978): 201–7.

Hess RF, Wang Y-Z, Demanins R, Wilkinson F, Wilson HR. A deficit in strabismic amblyopia for global shape detection. *Vision Research* 39 (1999): 901–14.

Howard IP, Rogers BJ. *Seeing in Depth.* Ontario: I. Porteus, 2002, ch. 13.

Levi DM, Song S, Pelli DG. Amblyopic reading is crowded. *Journal of Vision* 7 (2007): article 21, 1–17.

McKee SP, Levi DM, Movshon JA. The pattern of visual deficits in amblyopia. *Journal of Vision* 3 (2003): article 5, 380–405.

Simmers AJ, Ledgeway T, Mansouri B, Hutchisnon CV, Hess RF. The extent of the dorsal extra-striate deficit in amblyopia. *Vision Research* 46 (2006): 2571–80.

Sireteanu R, Baumer CC, Iftime A. Temporal instability in amblyopic vision: Relationship to a displacement map of visual space. *Investigative Ophthalmology and Visual Science* 49 (2008): 3940–54.

149   *To reduce amblyopia, you must train the eye and brain in many different tasks.*

Ciuffreda KJ, Levi DM, Selenow A. *Amblyopia: Basic and Clinical Aspects.* Boston: Butterworth-Heinemann, 1991.

Griffin JR, Grisham JD. *Binocular Anomalies: Diagnosis and Vision Therapy.* New York: Butterworth-Heinemann, 2002.

Krumholtz I, Fitzgerald D. Efficacy of treatment modalities in refractive amblyopia. *Journal of the American Optometric Association* 70 (1999): 399–404.

Press, LJ. *Applied Concepts in Vision Therapy.* St. Louis, MO: Mosby, 1997.

Wick B, Wingard M, Cotter S, Scheiman M. Anisometropic amblyopia: Is the patient ever too old to treat? *Optometry and Vision Science* 69 (1992): 866–78.

149   *In 2007, Robert Hess and his colleagues at McGill University published a study demonstrating that people with amblyopia do not discount all information from the amblyopic eye.*

Baker, DH, Meese TS, Mansouri B, Hess RF. Binocular summation of contrast remains intact in strabismic amblyopia. *Investigative Ophthalmology and Visual Science* 48 (2007): 5332–38.

150   *They have developed a class of procedures called "monocular fixation in a binocular field" (MFBF).*

Brock FW. New methods for testing binocular control. *Journal of the American Optometric Association* 34 (1963): 443–50.

Cohen AH. Monocular fixation in a binocular field. *Journal of the American Optometric Association* 52 (1981): 801–6.

Press LJ. *Applied Concepts in Vision Therapy.* St. Louis, MO: Mosby, 1997.

Schapero M. *Amblyopia.* Philadelphia: Clinton Book Co., 1971, 253–54.

151   *In one published study, one-third of the 203 amblyopes examined, many of whom were older adults, experienced significantly improved eyesight in their amblyopic eye after vision was lost in the fellow eye.*

Vereecken EP, Brabant P. Prognosis for vision in amblyopia after the loss of the good eye. *Archives of Ophthalmology* 102 (1984): 220–24.

Daw NW. Critical periods in the visual system. In Hopkins B, Johnson SP (eds.), *Neurobiology of Infant Vision.* Westport, CT: Praeger Pub., 2003, 64–67.

151   *As early as 1957, Carl Kupfer published a study in which he showed dramatic improvements in adult amblyopes after a four-week period of patching combined with vision therapy.*

Kupfer C. Treatment of amblyopia ex anopsia in adults: A preliminary report of seven cases. *American Journal of Ophthalmology* 43 (1957): 918–22.

152   *In a 1977 study, Martin Birnbaum and his colleagues reviewed twenty-three published studies on amblyopia and reported that improvements in eyesight were found for all ages.*

Birnbaum MH, Koslowe K, Sanet R. Success in amblyopia therapy as a function of age: A literature survey. *American Journal of Optometry and Physiological Optics* 54 (1977): 269–75.

In 1992, Bruce Wick and colleagues published a study in which dramatic improvements in acuity of the amblyopic eye as well as binocular vision were obtained in anisometropic amblyopes with the proper glasses, part-time occlusion, and vision therapy. This study contrasted with published results by Meyer and colleagues, who found that occlusion treatment alone produced dramatic improvements only in patients under ten years of age. The difference between these two studies was the use of vision therapy.

Wick B, Wingard M, Cotter S, Scheiman M. Anisometropic amblyopia: Is the patient ever too old to treat? *Optometry and Vision Science* 69 (1992): 866–78.

Meyer E, Mizrahi E, Perlman I. Amblyopia success index: A new method of quantitative assessment of treatment efficacy application in a study of 473 anisometropic amblyopic patients. *Binocular Vision Quarterly* 6 (1991): 75–82.

Other studies demonstrating successful treatment of older amblyopes are listed below:

Pediatric Eye Disease Investigator Group. Randomized trial of treatment of amblyopia in children aged 7 to 17 years. *Archives Ophthalmology* 123 (2005): 437–47.

Rutstein RP, Fuhr PS. Efficacy and stability of amblyopia therapy. *Optometry and Vision Science* 69 (1992): 747–54.

Saulles H. Treatment of refractive amblyopia in adults. *Journal of the American Optometric Association* 58 (1987): 959–60.

Selenow A, Cuiffreda KJ. Vision function recovery during orthoptic therapy in an adult esotropic amblyope. *Journal of the American Optometric Association* 57 (1986): 132–40.

152 *For example, exposing adult rats to enriched environments reverses the amblyopia produced by monocular deprivation occurring in infancy.*
Sale A, Maya Vetencourt JF, Medini P, Cenni MC, Baroncelli L, De Pasquale R, Maffei L. Environmental enrichment in adulthood promotes amblyopia recovery through a reduction of intracortical inhibition. *Nature Neuroscience* 10 (2007): 679–81.

152 *Placing adult rats in the dark for three to ten days can also reverse the effects of monocular deprivation.*
He H-Y, Hodos W, Quinlan EM. Visual deprivation reactivates rapid ocular dominance plasticity in adult visual cortex. *Journal of Neuroscience* 261 (2006): 2951–55.

He H-Y, Ray B, Dennis K, Quinlan EM. Experience-dependent recovery of vision following chronic deprivation amblyopia. *Nature Neuroscience* 10 (2007): 1134–36.

152 *Studies of "perceptual learning" in human amblyopes pioneered by Dennis Levi and Uri Polat in the mid-1990s demonstrate improvements in eyesight in the amblyopic eye even in adult patients.*

Fronius M, Cirina L, Kuhli C, Cordy A, Ohrloff C. Training the adult amblyopic eye with "perceptual learning" after vision loss in the non-amblyopic eye. *Strabismus* 14 (2006): 75–79.

Huang C-B, Zhou Y, Lu, Z-L. Broad bandwidth of perceptual learning in the visual system of adults with anisometropic amblyopia. *Proceedings of the National Academy of Sciences* 105 (2008): 4068–73.

Levi DM. Perceptual learning in adults with amblyopia: A reevaluation of critical periods in human vision. *Developmental Psychobiology* 46 (2005): 222–32.

Li RW, Klein SA, Levi DM. Prolonged perceptual learning of positional acuity in adult amblyopia: Perceptual template retuning dynamics. *Journal of Neuroscience* 28 (2008): 14223–29.

Polat U, Levi DM. Neural plasticity in adults with amblyopia. *Proceedings of the National Academy of Sciences* 93 (1996): 6830–34.

Polat U, Ma-Naim T, Belkin M, Sagi D. Improving vision in adult amblyopia by perceptual learning. *Proceedings of the National Academy of Sciences* 101 (2004): 6692–97.

Zhou Y, Huang C, Xu P, Tao L, Qiu Z, Li X, Lu Z-L. Perceptual learning improves contrast sensitivity and visual acuity in adults with anisometropic amblyopia. *Vision Research* 46 (2006): 739–50.

153 *In contrast, 38 to 50 percent of patients with infantile strabismus and 70 percent of patients whose strabismus develops after the first year acquire stereopsis through optometric vision therapy.*

Etting GL. Strabismus therapy in private practice: Cure rates after three months of therapy. *Journal of the American Optometric Association* 49 (1978): 1367–73.

Flax N, Duckman RH. Orthoptic treatment of strabismus. *Journal of the American Optometric Association* 49 (1978): 1353–61.

Ludlam WM. Orthoptic treatment of strabismus: A study of one hundred forty-nine non-operated, unselected, concomitant strabismus patients completing orthoptic training at the Optometric Center of New York. *American Journal of Optometry and Archives of the American Academy of Optometry* 38 (1961): 369–88.

Ludlam WM, Kleinman BI. The long range results of orthoptic treatment of strabismus. *American Journal of Optometry and Archives of the American Academy of Optometry* 42 (1965): 647–84.

Chapter 9: Vision and Revision

156 *In his fascinating book* Rebuilt: How Becoming Part Computer Made Me More Human, *Michael Chorost, who was hard of hearing as a young child and profoundly deaf in adulthood, describes how he learned to hear again using a cochlear implant.*
Chorost M. *Rebuilt: How Becoming Part Computer Made Me More Human.* New York: Houghton Mifflin Co., 2005, 126.

157 *Edward Taub and his colleagues at the University of Alabama discovered that the patients most actively involved in their own rehabilitation recovered best from strokes.*
Taub E, Uswatte G, Mark VW, Morris DM. The learned nonuse phenomenon: Implications for rehabilitation. *Europa Medicophysica* 42 (2006): 241–56.
Morris DM, Taub E, Mark VW. Constraint-induced movement therapy: Characterizing the intervention protocol. *Europa Medicophysica* 42 (2006): 257–68.

When a person suffers a stroke, he may lose the use of one of his limbs. Right after the stroke, the amount of brain area used to control, let's say, the affected arm has contracted. The person may have to exert great effort to move the arm, and the resulting movements are clumsy. So, he avoids use of the affected arm and finds other ways to move in order to compensate. With time and healing, he may actually regain better control of the weak arm, but it is too late. He has already learned to discount its use. Taub and his colleagues call this behavior "learned nonuse," and without intervention, this way of moving becomes permanent.

To counteract learned nonuse, Taub and his colleagues developed constraint-induced movement therapy (CI therapy) in which patients wear a restraining device, such as a mitt on the good arm, to force the use of the affected limb. During an intense two-week period, the patients also work in an office with a trained therapist or interventionist and practice arm-movement exercises of increasing difficulty. Taub and his colleagues assumed that these interventions would be the most important in helping their patients use the arm again. So, they were surprised to discover that the patients' ability to use their affected arm in everyday life did not correlate with their ability on motor performance tests. Instead, the most critical component of the therapy was a set of procedures called the transfer package, so named because it involved transferring the skills learned in therapy sessions to the real world. As part of the transfer package, the patients created a list of "activities of daily living" that involved the affected arm, such as tying shoes or eating with a spoon and fork. They then drew up a contract with their interventionist in which they stated how much they would use their affected arm in these activities. They kept a diary of how they were using their weakened limb and discussed with the interventionist how they solved problems involving the use of the arm. Taub writes, "A major difference in CI therapy is *the involvement of the patient as an active participant in all requirements of the therapy* not only during the treatment period but also (and especially) after laboratory therapy has been completed" (emphasis added).

158   *Indeed, as early as the mid-1900s, Frederick Brock stressed these ideas when he noted that to train strabismics to see with normal binocular vision and stereopsis, you have to challenge them with tasks that are close to, but just beyond, their current skills and with exercises that resemble actions experienced in real life.*

        Brock FW. Visual Training—part III. *Optometric Weekly* 46–50 (1955–59).

158   *If, however, these adults are forced to hunt, then they realign their brain's auditory spatial maps with the prism-altered visual maps.*

        Bergan JF, Ro P, Ro D, Knudsen EI. Hunting increases adaptive auditory map plasticity in adult barn owls. *Journal of Neuroscience* 25 (2005): 9816–20.

158   *The critical period encompasses the developmental stage when the brain changes in response to most strong stimuli, not just to behaviorally relevant ones.*

        This idea has been expressed by several investigators: Bao S, Chang EF, Davis JD, Gobeske KT, Merzenich MM. Progressive degradation and subsequent refinement of acoustic representations in the adult auditory cortex. *Journal of Neuroscience* 26 (2003): 10765–75.

        Keuroghlian AS, Knudsen EI. Adaptive auditory plasticity in developing and adult animals. *Progress in Neurobiology* 82 (2007): 109–21.

159   *As I suggest in chapter 6, some of the changes in my brain involved modifications of synaptic connections in the visual cortex, an area of the brain that is highly evolved in humans but much less developed in our distant vertebrate cousins.*

        All vertebrate animals have a cerebral cortex, which is an outgrowth of the roof of the forebrain. The cortex, however, takes on a whole new dimension in mammals. Many people learn in school that mammals can be distinguished from nonmammalian vertebrates by the presence of fur or hair for regulating body temperature and mammary glands for suckling the young. As John Allman points out in *Evolving Brains* (New York: Scientific American Library, 2000), another very important distinguishing characteristic of mammals is our many-layered cerebral cortex.

In nonmammalian vertebrates, including fish, amphibians, and reptiles, the cerebral cortex contains one to three layers of cells and is called the dorsal cortex. In contrast, in mammals, the cerebral cortex is made up of six principal layers, some with sublayers, and is called the neocortex.

159   *In fact, the wiring in the cerebral cortex reflects the history of our actions and is constantly reshaped by them.*

Polley DB, Steinberg EE, Merzenich MM. Perceptual learning directs auditory cortical map reorganization through top-down influences. *Journal of Neuroscience* 26 (2006): 4970–82.

The primary visual, primary auditory, and primary somatosensory cortices are the first areas of the cerebral cortex to receive input from the eyes, ears, and body, respectively. The circuitry of the primary sensory areas was once thought to develop early in life and then remain relatively immune to change. Recent research, however, indicates that this is not the case. For example, in experiments performed in the above-cited paper, adult rats were played a series of tones of different pitches and intensities. One group of rats was rewarded for correctly identifying a tone of a certain (five kilohertz) pitch. The loudness or softness of the five-kilohertz tone did not matter. In contrast, another group was rewarded for correctly identifying a tone of a certain intensity or loudness (twenty-five decibels), but the pitch of the tone did not matter.

Rats who were rewarded if they identified the five-kilohertz tone showed enhanced neuronal responses in their primary auditory cortices to tones of five kilohertz. As a result of their training, their auditory cortex was modified so that more neurons were responsive to the five-kilohertz tone. However, no changes in neuronal responses were seen to tones of different loudness levels. In contrast, rats who were trained to discriminate tones by their loudness showed enhanced neuronal responses to tones of the trained intensity but did not show changes in neuronal responses to pitch.

Particularly striking about these results is that both groups of rats were presented throughout the experiment with the *same* tones. The tones varied in pitch and intensity, and it was up to the rats to pick out the right ones for their reward. The tones themselves provided the external sensory input. The demands of the tasks shaped the relative importance of the input. So, the response of neurons in the brain did not change as a result of simply hearing the tones but instead according to what was important to the rat—to what the rats had to learn to get the reward.

Pleger B, Blankenburg F, Ruff CC, Driver J, Dolan RJ. Reward facilitates tactile judgments and modulates hemodynamic responses in human primary somatosensory cortex. *Journal of Neuroscience* 28 (2008): 8161–68.

In the above-cited study, changes with learning in the human primary somatosensory cortex were observed using functional MRI. Vibrations of two different frequencies were applied to the index finger of human volunteers, and the volunteers then reported which of the two stimuli was at the higher frequency. Prior to this test, the volunteers were shown on a computer screen how much money they would receive for the right answer. Their performance on this task improved with an increase in the size of the monetary reward. After each choice, the volunteers were given visual (but not somatosensory) feedback as to whether or not they had given the right answer and how much money they had earned. Not surprisingly, their somatosensory cortex, the part of the cortex that responds to touch, was active as the volunteers' fingers were vibrated. Intriguing was the fact that the somatosensory cortex reactivated when the volunteers received a visual signal indicating their reward. The investigators postulated that upon receipt of the reward, the somatosensory cortex was reactivated because it was reorganizing itself in a way that made it more sensitive to the relevant vibrations.

160 *In several different experiments, adult guinea pigs or rats were played a sound of a particular pitch, while neuronal activity was monitored in their auditory cortex.*

Bakin JS, Weinberger NM. Induction of a physiological memory in the cerebral cortex by stimulation of the nucleus basalis. *Proceedings of the National Academy of Sciences* 93 (1996): 11219–24.

Kilgard MP, Merzenich MM. Cortical map reorganization enabled by nucleus basalis activity. *Science* 279 (1998): 1714–18.

160 *Similar results have also been seen when presentation of a particular stimulus is paired with activation of other neuromodulatory areas of the brain.*

Bao S, Chan VT, Merzenich NM. Cortical remodeling induced by activity of ventral tegmental dopamine neurons. *Nature* 412 (2001): 79–83.

Bear MF, Singer W. Modulation of visual cortical plasticity by acetyl-choline and noradrenaline. *Nature* 320 (1986): 172–76.

Gu Q. Neuromodulatory transmitter systems in the cortex and their role in cortical plasticity. *Neuroscience* 111 (2002): 815–35.

Kasamatsu T, Watabe K, Heggelund P, Scholler E. Plasticity in cat visual cortex restored by electrical stimulation of the locus coeruleus. *Neuroscience Research* 2 (1985): 365–86.

Kentros CG, Agnihotri NT, Streater S, Hawkins RD, Kandel ER. Increased attention to spatial context increases both place field stability and spatial memory. *Neuron* 42 (2004): 283–95.

Kojic L, Gu Q, Douglas RM, Cynader MS. Serotonin facilitates synaptic plasticity in kitten visual cortex: An in vitro study. *Developmental Brain Research* 101 (1997): 299–304.

Li S, Cullen WK, Anwyl R, Rowan MJ. Dopamine-dependent facil-itation of LTP induction in hippocampal CA1 by exposure to spatial novelty. *Nature Neuroscience* 6 (2003): 526–31.

Weinberger NM. Associative representational plasticity in the audi-tory cortex: A synthesis of two disciplines. *Learning and Memory* 14 (2007): 1–16.

161   *The neuromodulators may have helped to unmask and strengthen con-nections that had been ineffective but were never entirely lost.*

Brocher S, Artola A, Singer W. Agonists of cholinergic and noradren-ergic receptors facilitate synergistically the induction of long-term poten-tiation in slices of rat visual cortex. *Brain Research* 573 (1992): 27–36.

Kasamatsu T, Pettigrew JD. Depletion of brain catecholamines: Fail-ure of ocular dominance shift after monocular occlusion in kittens. *Science* 194 (1976): 206–9.

Kirkwood A, Rozas C, Kirkwood J, Perez F, Bear MF. Modulation of long-term synaptic depression in visual cortex by acetylcholine and nor-epinephrine. *Journal of Neuroscience* 19 (1999): 1599–609.

161   *In addition, the same neuromodulators that triggered and facilitated synaptic changes may have made these changes long-lasting.*

Bailey CH, Giustetto M, Huang Y-Y, Hawkins RD, Kandel ER. Is heterosynaptic modulation essential for stabilizing Hebbian plasticity and memory? *Nature Reviews. Neuroscience* 1 (2000): 11–20.

Ma X, Suga N. Long-term cortical plasticity evoked by electric stim-ulation and acetylcholine applied to the auditory cortex. *Proceedings of the National Academy of Sciences* 102 (2005): 9335–40.

161 *High levels of neuronal activity in the brainstem and basal forebrain are seen in animals when they are alert and exploring their environment, when they are learning about novel stimuli, and when they anticipate a reward for their actions.*

Aston-Jones G, Rajkowski J, Kubiak P. Conditioned responses of monkey locus coeruleus neurons anticipate acquisition of discriminative behavior in a vigilance task. *Neuroscience* 80 (1997): 697–715.

Corbetta M, Patel G, Shulman GL. The reorienting system of the human brain: From environment to theory of mind. *Neuron* 58 (2008): 306–24.

Richardson RT, DeLong MR. Context-dependent responses of primate nucleus basalis neurons in a go/no-go task. *Journal of Neuroscience* 10 (1990): 2528–40.

162 *My surprising and delightful views sent me on hunts for new sights.*

Beverly Biderman expresses the same feelings in her book *Wired for Sound: A Journey into Hearing* (Toronto: Trifolium Books, 1998) when she describes going on "sound hunts" after receiving a cochlear implant.

162 *After training, neurons in the uninjured areas of the somatosensory cortex respond to touch from the affected fingers.*

Xerri C, Merzenich MM, Peterson BE, Jenkins W. Plasticity of primary somatosensory cortex paralleling sensorimotor skill recovery from stroke in adult monkeys. *Journal of Neurophysiology* 79 (1998): 2119–48.

163 *Not surprisingly, then, neurons have been found in the brain that are sensitive to both tactile and visual input and are involved in our ability to reach for and grasp objects.*

Colby CL, Goldberg ME. Space and attention in parietal cortex. *Annual Review of Neuroscience* 22 (1999): 319–49.

163 *The responses of some neurons in the visual cortex are modulated by sounds.*

Allman BL, Meredith MA. Multisensory processing in "unimodal" neurons: Cross-modal subthreshold auditory effects in cat extrastriate visual cortex. *Journal of Neurophysiology* 98 (2007): 545–49.]

Capp C, Barone P. Heteromodal connections supporting multisensory integration at low levels of cortical processing in the monkey. *European Journal of Neuroscience* 22 (2005): 2886–902.

Clemo HR, Sharma GK, Allman BL, Meredith MA. Auditory projections to extrastriate visual cortex: Connectional basis for multisensory processing in "unimodal" visual neurons. *Experimental Brain Research* 191 (2008): 37–47.

Wang Y, Simona C, Trotter Y, Barone P. Visuo-auditory interactions in the primary visual cortex of the behaving monkey: Electrophysiological evidence. *BMC Neuroscience* 9 (2008): 79. Available online at www.biomedcentral.com/1471-2202/9/79.

163 *The books* Privileged Hands *and* Touching the Rock *both highlight in fascinating detail what it is like to be blind.*

Vermeij G. *Privileged Hands: A Scientific Life*. New York: W. H. Freeman and Co., 1997.

Hull JM. *Touching the Rock: An Experience of Blindness*. New York: Pantheon Books, 1990.

163 *Indeed, imaging studies of the brains of individuals who are born blind or become blind in early life indicate that they use their visual cortex for nonvisual activities.*

Pascual-Leone A, Amedi A, Fregni F, Merabet LB. The plastic human brain cortex. *Annual Review of Neuroscience* 28 (2005): 377–401.

164 *However, the National Federation of the Blind has reported that audible traffic signals can actually compromise safety.*

Available online at http://access-board.gov/PROWAC/comments/comments-10–28/mccarthy-j.htm.

164 *In Spain, future teachers of the blind must spend a week with blindfolds on in an attempt to experience firsthand what it is like not to see.*

Pascual-Leone A, Amedi A, Fregni F, Merabet LB. The plastic human brain cortex. *Annual Review of Neuroscience* 28 (2005): 377–401.

164 *Alavaro Pascual-Leone and his colleagues at Harvard University blindfolded sighted individuals for one week and imaged their brains while the participants attempted to identify braille letters or the pitch of a sound.*

Pascual-Leone A, Amedi A, Fregni F, Merabet LB. The plastic human brain cortex. *Annual Review of Neuroscience* 28 (2005): 377–401.

Saito DN, Okada T, Honda M, Yonekura Y, Sadato N. Practice makes perfect: The neural substrates of tactile discrimination by Mah-Jong experts include the primary visual cortex. *BBMC Neuroscience* 7 (2006): 79. Available online at http://biomedcentral.com/1471–2202/7/79.

Saito and colleagues have also shown that tactile stimulation can activate the primary visual cortex in normally sighted people. Their study involved participants who were experts at the game Mah-Jong and could readily identify by feel the carved marks in a Mah-Jong tile. When these players felt the tiles with their eyes closed, their primary visual cortex and lateral occipital cortex were activated. Visual cortical areas were not activated by the same task in control observers who did not play Mah-Jong.

165 *When I consider the extraordinary changes produced in the blindfold study described above, the alterations that may have occurred in my brain do not seem so surprising.*

The visual cortex and neuromodulatory areas may not be the only brain areas involved in my vision changes. The visual cortex has a continual back-and-forth dialogue with a brain region called the thalamus and indirectly the retina, and changes may have occurred in these pathways as well. To learn to see in stereo, I had to concentrate hard on the therapy tasks at hand, and focused attention requires input from nonvisual parts of my cerebral cortex. In fact, imaging studies on humans reveal that a different pattern of whole-brain activity is seen when we focus on a demanding task than when we perform an automatic action.

Corbetta M, Patel G, Shulman GL. The reorienting system of the human brain: From environment to theory of mind. *Neuron* 58 (2008): 306–24.

Fox MD, Corbetta M, Snyder AZ, Vincent JL, Raichle ME. Spontaneous neuronal activity distinguishes human dorsal and ventral attention systems. *Proceedings of the National Academy of Sciences* 103 (2006): 10046–51.

Fox MD, Snyder AZ, Vincent JL, Corbetta M, Van Essen DC, Raichle ME. The human brain is intrinsically organized into dynamic, anticorrelated functional networks. *Proceedings of the National Academy of Sciences* 102 (2005): 9673–78.

For me, stereovision came on gradually, allowing me to see objects close to me, then more distant objects, in three dimensions. For Tracy Gray, the woman with torticollis described in chapter 4, both stereovision and an increase in peripheral vision appeared abruptly and in full measure.

Tracy states, "I was sitting on the sofa in my living room about midnight, and I immediately saw the whole room and all the objects in it in 3D. The breadth of it was not just the 3D vision, but the fact that I was able to take in so much more of the room than I did before." Jennifer Clark, who wrote to me from England, described a very similar experience. Accounts like Tracy's and Jennifer's suggest a sudden and global change in brain state, a change in the activity of whole populations of neurons.

Index

Allman, John, 231
Accommodation, 205
Accommodative esotropia, 23, 30, 113, 171
Acetylcholine, 160, 233–234
Adaptation, to disorders, 55–60
ADHD. *See* Attention deficit hyperactivity disorder
Alternating esotropia, 18, 171
Alternating fixation, 33, 40
Alvarez, Bruce, 54
Amblyopia (lazy eye; wandering eye), 52, 128–129, 138, 148, 171
   and causes of, 147–8, 222–223
   and critical periods (in visual development), 10–11, 145, 146–147, 151, 222
   and effect of age on recovery from, 151–153, 226–227
   and motion parallax, 126–127, 213
   and nature-versus-nurture debate, 145–153
   and occlusion therapy (patching), 148–149, 224
   and perceptual learning, 152–153, 182, 227–228
   and recovery from amblyopia after loss of non-amblyopic eye, 52, 195, 226
   and vision therapy, 149–152, 225–227
Amphibians, 136
Anisometropia, 148, 171
Anisometropic amblyopia, 148, 223, 227
Anomalous correspondence, 54–55, 171, 190, 195, 221
Archimedes, 4
Arnheim, Rudolf, 213
Artists, 107–111, 211–212
Astronauts, xi, 12–14, 59, 75–76, 77, 81, 181, 201
Atmospheric perspective, 110, 110 (fig.)
Atropine, 149
Attention deficit hyperactivity disorder (ADHD), 40–42
Audible traffic signals, 164
Auditory cortex, 159, 160, 231, 232, 233, 234. *See also* Cerebral cortex; Somatosensory cortex; Visual cortex
Awareness, 157

Bach-y-Rita, Paul, 70, 199–200

Balance boards, 78–79, 201

Barn owl, 158, 230

Barry, Andy, 132

Barry, Dan, 55–56, 58–59, 65, 81, 84, 86, 102
    as astronaut, xi, 12–14, 81

Barry, Jenny, 12–13, 14, 127, 132

Barry, Susan, 19 (photo)
    and bicycling, 43, 83–84
    and blindness, fear of, 32
    and crew team, 44
    and distance viewing, 58–59, 64
    and driving, 43–44, 46, 58–59, 64–65, 78, 80, 84–85
    and eye fatigue, 64–65
    and floating unmoored, sensation of, 128
    and gaining stereopsis, doubts about, 98, 101, 133
    and glasses, first pair of, 29–30
    and glasses, with prism, 66, 67
    and glasses, without prism, 166
    and heights, fear of, 128
    and sense of immersion in space, 122–128
    and jittery vision, 46, 58, 67
    and latent stereovision, 142
    and mirrors, 120
    and movies, 80, 102–103, 127
    and music, 84, 130–131, 202
    and National Public Radio, interview on, 112, 209
    and object location, 65, 125
    and optometric records, 198–199, 213–214
    and ornithology, 44–45
    and piano, 130–131, 202
    and reading, 35–40, 64
    school photos of, 32–33
    and sensory overload, 131
    and sewing, 43
    and stereoblindness, first learning of, xiii, 1–4
    and stereomicroscope, use of, 45
    and stereopsis, acquisition of, xii–xiii, xiv–xv, 88, 89–90, 94–103, 105–107, 110–112, 117–118, 119, 120–123, 124–126, 127–128, 130–132, 162, 214
    and stereopsis, preoccupation with, 98, 131
    and strabismus as adult, xi–xiii, 44–45, 46, 59–60, 80
    and strabismus as child, 2, 11, 17–20, 35–40, 43
    and strabismus as infant, xi, 2, 11, 17–20, 23, 24–26, 28
    and surgery (eye), xi, 30–34, 61, 101, 145
    and tennis, 65, 78
    and thought process, change in, 131–132
    and tidiness, need for, 81–82
    and vision therapy, 60–61, 64, 67, 70–88, 89–94, 117, 120, 155
    and walking and running, 43
    and window views, 120

Basal forebrain, 160, 161, 233

Basketball, 79, 80

Bates method, 62

Begley, Sharon, 182

Behavioral optometrists, 34, 150, 172. *See also* Optometrists; Developmental optometrists

Bell, Mrs., 36

Berthoz, Alain, 200

Bicycling, 43, 83–84

Biderman, Beverly, 234

Bifocals, 30, 113

Binocular continuum, 144, 220

Binocular depth neurons, 138, 140–141, 218–219. *See also* Binocular neurons; Neurons

Binocular neurons, 3, 11, 62, 100 (fig.), 101, 138, 140–141, 143–144, 179, 216, 218–219. *See also* Binocular depth neurons; Neurons

Binocular rivalry, 22, 23, 185

Binocular vision, 3, 6, 24, 62–63, 67, 94, 123, 125, 142–143, 144, 158

Binocular vision disorders, 29, 43 permanence of, 10–11, 133–139, 147 (*see also* Critical periods)

Birch, Eileen E., 141, 184, 188, 189, 196, 197, 200, 219, 222, 224

Bird, Larry, 79

Birnbaum, Martin, 152, 226

Black Stallion series, 37

Blakemore, Colin, 207, 208, 220, 222, 223, 224

Blakeslee, Sandra, 102, 208

Blind spot, 6 (fig.), 40

Blindspot mechanism, 192

Blindness, 47, 163–165, 235 fear of, 32

Body movement and vision, 47–49. *See also* Movement

Bradley, Bill, 79

Braille, 11, 14, 163–164, 164–165, 181, 235

Brain and Visual Perception (Hubel and Wiesel), 140, 179, 196 216, 218, 223

Brain injury, 64

Brain plasticity, xv, xvi, 11–15, 54, 153, 158, 159–165, 181–182, 206–208, 226–228, 229–236

and synaptic connections, 98–101, 100 (fig.), 158–165, 206, 207–208, 233–235 *See also* Auditory cortex; Learning, neuronal basis of; Neural development; Neuromodulators; Neurons; Somatosenosory cortex; Synaptic connections; Visual cortex

Brainstem, 159–160, 161, 233

Brock, Frederick W., 54–55, 63, 64, 90, 93–94, 103, 117–118, 143, 158, 186, 194, 195, 196, 203, 204–206, 208, 209, 211, 220, 226, 230

Brock string, 90–94, 91 (fig.), 94–95, 97, 99, 101, 157, 161–162, 166, 204–206,

Brock training techniques, 90–94, 158, 204–206

Broglie, Princess Albert de, 105–106, 106 (photo), 108

Brown, Garry, 193

Brown, Stephanie Willen, 112–113

Burian, Hermann M., 203, 209

Cataracts, 56–57, 114, 146, 182, 222

Cats, xiii, 1–2, 24, 137, 137 (fig.), 138–139, 140, 146, 179, 181, 182, 186, 196, 207, 208, 216, 217, 221, 222, 223, 233, 234

Cerebral cortex, 159, 160, 231–232. *See also* Auditory cortex, Somatosensory cortex, Visual cortex

Chavasse, Francis Bernard, 135, 139, 215

Children
and eye surgery, 34, 62–63, 101,
135, 139, 153, 189, 190, 196–
197
See also Infants
Chorost, Michael, 156, 229
Clark, Jennifer, 129–130, 205–206,
236
Cochlear implant, 156–157
Cole, Eliza, 78–79
Coleman, Nick, 211
College of Optometrists in Vision
Development (COVD), 64,
175, 198
Colorblindness, 10, 180
Columbia School of Optometry, 90
Compensatory eye movements, 76–
77, 201
Concentration, 157, 236
Contact lenses, 63, 152, 198
"Contributions to the Physiology of
Vision–Part the First. On some
remarkable, and hitherto
unobserved, Phenomena of
Binocular Vision"
(Wheatstone), 6, 180
Controlled reading, 38. *See also*
Reading
Convergence, 5, 5 (fig.), 20, 23, 42,
89, 93, 113, 115, 135, 172,
184, 185, 192–193, 203, 204–
206. *See also* Convergence
insufficiency; Divergence;
Vergence eye movements
Convergence insufficiency, 42–43,
172, 192–193, 205–206. *See
also* Convergence
Cooper, Rachel, 123–124, 176, 211
Corbit, Margaret, 211–212
Cornea, 6 (fig.)

Correlated input from the two eyes,
99–101, 206–207
Corresponding retinal regions, 6, 50,
172, 194–195
Corridor illusion, 119–120, 119
(fig.)
Costenbader, Frank D., 184, 217
COVD. *See* College of Optometrists
in Vision Development
Crew team, 44
Critical periods (in visual
development), xiii, 2, 10–11,
15, 62, 63, 140, 153–154,
158–159, 181–183, 196, 222
and amblyopia, 10–11, 145, 146–
147, 151
new views about, xv–xvi, 158–
159, 230
and strabismus, 10–11, 139–140,
218
Crone, Robert A., 179
Crossed eyes, 172. *See* Esotropia
Cross-fixate, 40. *See also* Fixate
Crowding in strabismus and
amblyopia, 210

Danner, Mrs., 36–37
Daw Nigel W., 185, 186, 194, 196,
200, 222, 223, 225, 226
Deafness, 156–157
Deprivation amblyopia, 147, 222
Depth perception, 20, 28–29, 33,
124, 125–127, 172, 185, 186,
188, 204, 205–206, 211, 212–
213
in art, 107–110, 209, 211–212
Developmental optometrists, xii, 34,
40, 42, 60, 133, 145, 147, 150,
152, 172, 175, 198. *See also*
Behavioral optometrists;
Optometrists

Dilts, Michelle, 71
Disorders, adaptation to, 55–60
Distance and space, sense of, 29,
    102–103, 134, 172, 203. *See
    also* Depth perception.
Divergence, 5, 5 (fig.), 27, 39, 89,
    93, 115, 135, 172, 203. *See also*
    Convergence; Vergence eye
    movements
Doidge, Norman, 182
Dolezal, Hubert, 80, 202
Dopamine, 160, 233
Dore, Eric, 40–43
Dore, Michelle, 40–42
Double vision, 24, 25 (fig.), 26–28,
    79
Driving, 26, 43–44, 46, 55, 58–59,
    64–65, 78, 80, 84–85
Duane's syndrome, 133, 214
Duffy, Pat, 144–145, 203
Duke-Elder, Sir Stewart, 189, 197
Dynamic visual acuity testing, 75–
    76, 201

Ear, 76–77
Escher, M. C., 109
Esotropia (crossed eyes), 18, 23, 172
    accommodative, 23, 30, 113
    alternating, 18
    infantile, 23, 173, 184, 185, 188,
        189–190, 200, 217–218
Euclid, 4
Exotropia (walleye), 18, 172
Eye alignment test (cover test), 19,
    155–156
Eye fatigue, 64–65
Eye movements, See Convergence,
    Divergence, Saccades, Smooth
    pursuits, Vergence eye
    movements

Fahle, Manfred, 182
Far lookers, 81
Farley, Walter, 37
Farsightedness, 113, 171
Fasanella, Rocko, 18–20, 28, 29,
    30–31, 32, 33
Feinstein, Malcolm, 110, 111
Fish, 136
Fitzpatrick, Heather, 83 (photo),
    124, 161–162
Fitzpatrick, Tara, 81, 210–211
Fixate, 5 (fig.), 18, 72–73, 173. *See
    also* Cross–fixate; Gaze holding
Floating unmoored, sensation of,
    128
Four-corners exercise, 71, 73
Fovea, 6, 6 (fig.), 20, 21, 25, 27, 28,
    40, 49, 50, 53–54, 89, 91, 138,
    173
Francke, Amiel, 58, 78, 201
Free-fuse, 10, 121. *See also* Fusion
Frogs, 136
Frost-Arnold, Greg, 180
Fully adapted strabismics, 54
Fusion, xvi, 4, 7, 8, 10, 21, 23–24,
    34, 91–92, 115, 117–118, 118
    (fig.), 119, 121, 125, 135, 142,
    143, 145, 166, 185, 186, 190,
    199, 204, 205–206, 209, 214,
    215. *See also* Panum's fusional
    area; Peripheral fusion;
    Stereopsis
Fusion effort, 94. *See also* Fusion

Gait, 29
Garzia, Ralph Philip, 180, 195
Gaze, 67, 72–78, 166. *See also* Gaze
    holding
Gaze holding, 71–72, 74–75, 200
    and movement, 75–78
    See also Fixate

Geese, 136, 215–216
Gesell, Arnold, 186
Gesell Institute of Human
  Development, 34
Gibson, James J. 194, 199, 204
Goethe, Johann Wolfgang von, 4
Goldstein, Kurt, 55, 195
Gray, Tracy, 57–58, 129, 130, 196,
  236
Graylag geese, 136, 215–216
"The Green World" (Hochman),
  114
Greene, Red (nickname), 52–53,
  151
Greenwald, Israel, 64, 204
Gregory, Richard L., 182, 209
Gretsky, Wayne, 79
Griffin, John R., 191, 193, 197. 225
Grisham, J. David, 191, 193, 197,
  225
Gruning, Carl, 124

"Hands" (Sacks), 96–97, 112, 206
Harmon, Darell Boyd, 194, 196
Harris, Paul, 82, 144, 220
Harvard, 11, 136, 164
Hearing, 156–157
Hebb, Donald, 206
Heights, fear of, 128
Held, Richard 22, 185, 186, 187,
  219
Helveston, Eugene M. 184, 191, 217
Hering, Ewald, 134, 215
Hess, Robert, 149–150, 225–226
Hillier, Carl, 129
Hochman, Rachel, 56–57, 114–117
Hole-in-the-hand experiment, 51–
  52, 51 (fig.), 54
Hong, Carole, 152
Horopter, 180
Howard, Ian P., 179, 195, 215, 225

Hubel, David, xiii, 11, 136–140,
  146–147, 148, 179, 181, 196,
  200, 216, 217, 218, 219, 221–
  222, 223
Hull, John, 47, 163, 194, 235
Human eye, 6, 6 (fig.)
  muscles of, 30 (fig.)
Hurst, Caroline, 129, 205
Hyperactivity disorder, 40–42

Immersion in space, sense of, 122–
  128
Imprinting, 136, 215–216
Infantile esotropia, 23, 173, 184,
  185, 188, 189, 190, 200, 217–
  218
Infants, 20–24, 28–29, 134, 135,
  140–142, 153
  and eye surgery, 34, 62–63, 101,
    135, 139, 153, 189, 190, 196–
    197
  and preferential-looking
    experiments, 21–23, 22 (fig.).
    141, 185, 219
  and seeing in 3D, 21–23, 185,
    219
Ingres, Jean Auguste Dominique,
  105, 107, 108
Inhibitory connections between the
  pathways of the two eyes, 101,
  207–208

Jackson, Frank, 180
James, William, 20
Jampolsky, Arthur, 202
Javal, Louis Emile, 62, 73, 197
Jittery vision, 46, 58
Johns Hopkins University, 136
Johnson Space Center (NASA), 75–76
Julesz, Bela, 209–210, 219

Kandel, Eric, R., 206, 233, 234
Kim, Julie, 152
Kiorpes, Lynne, 148, 223, 225
Knudsen, Eric, 182, 230
Kraskin, Robert, 78
Krulwich, Robert, 206
Kupfer, Carl, 151–152, 226
Kurson, Robert, 183

Latent nystagmus, 73, 183, 199, 200
Lateral geniculate nucleus, 185
Latent stereovision, 142–145
Lazy eye, 173. *See* Amblyopia
Learning,
    and neuronal basis of, 98–
        101,158–162, 206–208,
        229–236
    and vision, 35–43, 191–193
Learning disorders, 40–42
Learned nonuse, 229–230
Leclerc, Georges-Louis, Comte de
    Buffon, 148
Legally Blonde (film), 127
Leigh, John R., 200
Lens (human eye), 6 (fig.), 21
Lessmann, Hans, 57, 114–117
Lester, Joan, 152
Levi, Dennis M., 152–153, 182,
    210, 222, 227–228
Living in a World Transformed
    (Dolezal), 80, 202
Livingstone, Margaret S., 209
Loading, 87–88
Long-term potentiation, 99–101,
    206–207
Looking-soft technique, 83–84,
Lorenz, Konrad, 136, 137, 215–216
Low-vision devices, 63
Ludlam, William, 63, 183, 198, 221,
    228
Lundin, Margaret, 203

Macaque monkeys, 140, 162, 184,
    187–188, 200, 218, 219, 222,
    223, 233, 234, 235. *See also*
    Monkeys
Macula, 6 (fig.)
Madeleine J., 96–97
Magic Eye books, 121–122
Magnetic source imaging, 11–12,
    181
Mary the Neuroscientist, 180
*The Man Who Mistook His Wife for a
    Hat* (Sacks), 96, 112, 206
Markow, Steven, 60
Massachusetts Institute of
    Technology (MIT), 21, 24,
    185, 186
McGill University, 149
McPhee, John, 202
Merhar, Sarah, 26–28,186, 203
Merleau-Ponty, Maurice, 131
Merzenich, Michael M. 182, 230,
    231–232, 234
MFBF. *See* Monocular fixation in a
    binocular field
Milstein, Nathan, 84
Mirrors, 120
MIT. *See* Massachusetts Institute of
    Technology
Mitchell, Donald E., 204, 207, 224
Monet, Claude, 105, 123
Monkeys, 137, 140, 146–147, 148,
    162, 184, 187–188, 200, 218,
    219, 222, 223, 233, 234, 235
Monocular brain, 4, 138
Monocular deprivation experiments,
    146–147, 148, 221–222, 223
Monocular fixation in a binocular
    field (MFBF), 150, 226
Monocular neurons, 3, 4, 138, 146,
    179, 181, 216. *See also* Neurons
Monter, Cyndi, 128–129, 147

Motion parallax, xi, 29, 125–127, 188, 212, 213
Movement
and gaze holding, 75–78, 200
and perception, 69–70, 90–94, 96, 199–200, 204–206
See also Body movement and vision
Movies, 80, 102–103, 127
Movshon, J. Anthony, 148, 192, 223, 224, 225
Multisensory neurons, 163, 234–235
Music, 84, 130–131, 202, 211, 213

NASA, 81
and Johnson Space Center, 75–76
National Eye Institute, 42–43, 193
National Federation of the Blind, 164, 235
National Public Radio, 112, 209
Nature-versus-nurture debate, 134–145, 153–154, 215–224
and amblyopia, 145–153, 222–228
and strabismus, 134–145, 215–221
Nawrot, Mark, 126, 188, 213
Near lookers, 81
Neural development, 134, 135–136, 137–139, 140–141, 143–144, 146, 148, 215–224. See also Brain plasticity; Neuromodulators; Neurons; Synaptic connections
Neuromodulators, 160–162, 232–234. See also Brain plasticity; Learning, neuronal basis of; Neurons; Synaptic connections
Neuromuscular disorders, 55–56
Neuronal plasticity, 54. See also Brain plasticity; Learning, neuronal

basis of; Neural Development; Neuromodulators; Neurons; Synaptic connections
Neurons, 3, 4, 62, 98–101, 100 (fig.), 134, 135, 158, 159, 160–161, 162–163, 181, 206–208, 216–219, 220, 222, 227, 231–236. See also Binocular neurons; Binocular depth neurons; Brain plasticity; Learnng, neuronal basis of; Neural development; Neuromodulators; Synaptic connections
New England Eye Center, Tufts University, 52
New York University, 148
New Yorker, The,112
Newton, Sir Isaac, 4
Nobel Prize, 149
Noe, Alva, 69, 194
Non-random dot stereogram, 121, 121 (fig.)
Noradrenaline, 160, 233–234
Norepinephrine, 160, 233–234
Normal correspondence, 51, 173
North Dakota State University, 126

Object location, 24, 49–55, 65, 125, 211
Object occlusion, in art, 109, 109 (fig.), 111
Occlusion, xi, 148–149, 224
Occulsion (patching) therapy. See Amblyopia
Ocular dominance columns, 185, 219, 222, 227
Ogle, Kenneth N. 204, 211
Ophthalmologists, 61–62, 103
and eye surgery, 62–63, 196–197
Ophthalmoscope, 134
Optic flow, 84–85

Optic nerve, 6 (fig.)
Optical illusion, 109, 109 (fig.), 119, 119 (fig.)
Optokinetic nystagmus, 183, 201
Optometrists, 15, 61, 133, 145
  and lens development, 63, 198
  and vision therapy, 62, 63–64, 197–198
  See also Behavioral optometrists; Developmental optometrists
The Organism (Goldstein), 55, 195
Ornithology, 44–45
Orthoptics, 62, 183, 198, 221, 228
Otolith organs, 76–77
Owl, 158, 230

Panum's fusional area, 92, 92 (fig.), 125, 173, 204. See also Fusion
Pascual-Leone, Alvaro, 164–165, 181, 235
Patching, 148, 149, 224. See also Amblyopia.
Pediatric Eye Disease Investigator Group, 149, 184, 227
Penneys, Rebecca, 213
Perception, and movement, 69–70, 90–94, 96, 199–200, 204–6
Perceptual learning, 152–153, 158, 182, 227–228
Perez, Steve, 142
Peripheral fusion, 115, 209. See also Fusion
Peripheral vision, 79–85
Perspective,
  and depth perception, xi, 2, 4–5, 9, 29, 102
  and art, 110, 111, 111 (fig.), 212
Phantoms in the Brain (Ramachandran and Blakeslee), 102, 208
Pharmacology, 62, 148–149

Phoropter, 66–67, 66 (fig.)
Physiological diplopia, 27
Piano, 130, 202, 213
Poggio, Tomaso, 182
Polaroid glasses, 21,115, 117, 143
Polaroid vectogram, 115–118, 116 (fig.), 118 (fig.), 119, 120, 166
Polat, Uri, 153, 227–228
Postsynaptic cell, 99–100, 100 (fig.)
Postsynaptic neuron, 99–100
Posture, 29, 56–58, 64, 196
Preferential-looking experiments, 21–23, 22 (fig.), 141, 185, 219
Press, Leonard J., 183, 193, 197, 198, 203, 209, 221, 225, 226
Prisms, 53, 63, 66, 67, 87–88, 87 (fig.), 166
Privileged Hands (Vermeij), 163–164, 235
Ptolemy, 195
Pupil (human eye), 6 (fig.)

Quale (qualia), 101–102, 208

Ramachandran, V. S., 102, 194, 195, 208, 219
Random dot stereogram, 121–122, 122 (fig.), 209–210
Reading, 35–40, 41–42, 64, 191, 192
Rebuilt: How Becoming Part Computer Made Me More Human (Chorost), 156, 229
Red and green panels, 86–88
Retina, 3, 6, 6 (fig.), 8–9, 20, 22, 27, 28, 40, 49, 50, 51, 53, 54, 71, 89, 91–92, 101, 107, 118, 125, 134, 138, 173
Retina Foundation of the Southwest, 141
Retinal disparity, 119, 120, 125, 140

Rogers, Brian J., 179, 195, 215, 225
Rominger, Kent, 59
Ronayne, Caitlin, 145–146, 147,
    149, 150–151
Rope circle (quoit) vectogram, 115–
    118, 116 (fig.), 118 (fig.), 119,
    120, 166
Rosenbaum, David A., 199, 204
Ruggiero, Theresa, 60–61, 62, 64–
    67, 70, 71, 72–73, 73–74, 77,
    82, 85, 86, 88, 90, 94, 98, 113,
    133, 134, 144, 145, 149, 150,
    154, 155–156, 166, 186, 198–
    199, 203, 213–214

Saccades, 73, 200
Sacks, Oliver, 96–97, 112, 180, 182,
    206, 209
Sadowski, Laurie, 87, 90–91, 93
Scheiman, Mitchell, 193, 209, 225,
    226–227
Scully, Lucas, 113–114
See Clearly method, 62
Semicircular canals, 76–77
Sengpiel, Frank, 207, 208, 220, 224
Sensory overload, 131
Serotonin, 160, 233
Sewing, 43
Shading, xi, 29, 102
    in art, 108, 108 (fig.), 111
Shadow, in art, 108–109, 109 (fig.),
    111
Shimojo, Shinsuke, 22, 185, 219
SILO phenomenon. See Small in,
    large out phenomenon
Sinha, Pawan, 182
Size constancy, 118–119
Skeffington, A. M., 78, 201
Slater, Alan, 184
Small in, large out (SILO)
    phenomenon, 118, 120, 209

Smooth pursuits, 73–74, 76, 126,
    188, 200–201, 212
Somatosensory cortex, 159, 162,
    163, 231, 232, 234. See also
    Auditory Cortex, Cerebral
    Cortex, Visual cortex
Spain, 164
Spalding, Douglas Alexander, 136,
    215–216
Spatial perception, See Depth
    Perception; Distance and Space,
    sense of; Object location
Sperry, Roger, 135–136, 215–216
Squint, 135, 173. See also Strabismus
Squint: Its Causes, Pathology, and
    Treatment (Worth), 135, 215
St. Benedict's Hospital, 96–97
St. Ours, Liz, 82–83
Stager, David, 141, 184, 188, 189,
    196, 197, 200, 219, 222, 224
Star Wars: Revenge of the Sith (film),
    102–103
Stasko, Kristina, 152
Steinman, Barbara A.,180, 195
Steinman, Scott B. 180, 195
"Stereo Sue" (Sacks), xv, 112, 206,
    209
Stereoblindness, xiii, 1–4, 123
Stereomicroscope, 45
Stereopsis, xii, xiv–xv, xv–xvi, 3, 4–9,
    24, 62, 63 67, 94, 101–102,
    107, 120–122, 123, 139, 142–
    143, 144, 158, 166, 173, 179,
    180, 185, 188, 190, 192, 203,
    204, 205–206, 209–210, 211,
    212, 213, 218–221
    explanation of sensation of, 102–
    103, 123–125
    onset of, in adults, 111, 112–117,
    124, 128–131
    qualitative, 211

Stereoscope, xiv, 6–10, 8 (fig.), 9
    (fig.), 70, 158, 205
Stereoscopy. *See* Stereopsis
Stereovision. *See* Stereopsis
Stern, Cathy, 186
Strabismic amblyopia, 147–148,
    222–223
Strabismic posture, 205
Strabismus (squint; misaligned eyes),
    2–3, 10–11, 173
  causes of, 23–24
  critical period for, 10–11, 138–
    140, 218
  family history of, 24, 221
  permanence of, 10–11, 133–140,
    166 (*see also* Critical periods)
  prevalence of, 18
  role of eye muscles in, 19, 184
  surgery for, xi, 30–34, 61, 62–63,
    101, 135, 139, 145, 153, 189,
    190, 196, 197
  types of, 23
  vision therapy for, 34, 62–64, 69–
    103, 117–118, 183, 198, 221,
    228
Stress, effects on vision, 155–156
Stroke, 53, 64, 157, 158–159, 162,
    165, 229–230, 234
Structure from motion, 127, 213
Suppression, interocular, 2–3, 26,
    28, 34, 85–88, 92–93, 187,
    188, 190, 202–203
Surgery (eye), xi, 30–34, 61, 62,
    101, 135, 139, 145, 153, 189,
    190, 196, 197
  on infants, 62–63, 153
Synaptic connections, 98–101, 100
    (fig.), 158–165, 206, 207–208,
    233–235. *See also* Brain
    plasticity; Learning, neuronal

basis of; Neural development;
    Neuromodulators; Neurons.

Taub, Edward, 157, 181, 229–230
Tennis, 65, 78
3D, 6, 9. *See also* Stereopsis
  and artists, 107–111, 211–212
  and infants, 21–23, 185, 219
3D cameras, 9
3D movies, 9
Torticollis, 58
Touching the Rock (Hull), 47, 163,
    194, 235
Traumatic brain injury, 64
Tufts University, 52
Tunicates, 48, 48 (fig.)
Tychsen, Lawrence, 188, 197, 200,
    201

University of Alabama, 157

Vergence eye movements, 5, 5 (fig.),
    20–21, 23, 89–94, 184, 185,
    192–193, 203, 204–206. *See
    also* Convergence; Divergence
Vermeij, Geerat, 163–164, 235
Vestibular organs, 76–78, 188, 201
Vestibulo-ocular reflex, 76–78, 188,
    201
View-Master, 9–10
Vinci, Leonardo da, 4
Violinists, 11–12, 14, 181
Vision
  and body movement, 47–49, 194
  and learning, 38–39, 191–193
  and learning disorders, 40–42,
    191–193
  and posture, 29, 56–58, 196
  and reading, 35–40, 41–42, 64,
    191, 192

Vision therapy, xvi, 57, 60–61, 62,
    63–64, 67, 70–88, 89–94,
    114–120, 153, 155, 193
  and amblyopia, 149–152, 225–
    227
  and awareness and concentration,
    157
  and balance boards, 78–79, 201
  and Brock string, 90–94, 91 (fig.),
    94–95, 97, 99, 101, 157, 161–
    162, 166, 204
  and convergence insufficiency,
    42–43, 192–193
  and eye movements, 72–74
  and four-corners exercise, 71, 73
  and gaze holding, 71–75
  and gaze holding and movement,
    75–78
  and infants, 197
  and latent stereovision, 142–145,
    220
  and loading, 87–88
  and looking-soft technique, 83–
    84, 202
  and optic flow, 84–85
  and peripheral vision, 79–85,
    201–202
  and prisms, 87–88
  and red and green panels, 86–88
  and relevance of procedures to real
    life activities, 157–158
  and rope circle (quoit) vectogram,
    115–118, 116 (fig.), 118 (fig.),
    119, 120, 166
  and saccades, 73
  and smooth pursuits, 73–74, 126
  and strabismus, 34, 62–64, 69–
    103, 117–118, 183, 198, 221,
    228

  and suppression, 85–88, 92
  and vergence movements, 85–94,
    204–206
  and wall game, 82–83, 150
Visual circuitry. See Learning,
    neuronal basis of; Synaptic
    connections, Visual cortex
Visual confusion, 24–28, 25 (fig.)
Visual cortex, xvi, 3–4, 11, 15, 140,
    144, 159, 220, 230–231
  and blindness, 163–164, 164–
    165, 235–236
  in cats, 137, 138–139, 146, 179,
    181, 182, 196, 207, 208, 216,
    217, 221, 222, 223, 233, 234
  changes in, 98–101, 160–161,
    206–207, 227, 232–234
  and connections to touch and
    hearing systems, 163–165;
    235–236
  in monkeys, 146, 148, 184, 187–
    188, 200, 218, 219, 222, 223
  See also Auditory cortex; Brain
    plasticity; Cerebral cortex;
    Learning, neuronal basis of;
    Neural development;
    Neuromodulators;
    Somatosensory cortex; Synaptic
    connections
Visual cortical neurons. See Visual
    cortex; Neurons
Visual direction, 50–55, 194, 195
Von Helmholtz, Hermann, 134, 215
Von Senden, Marius, 182
Von Noorden, Gunter K., 186, 192,
    195, 196, 210, 214, 215

Wald, Florence, 31
Waldman, Oliver, 193

Wall game, 82–83, 83 (fig.), 150
Walleye, 173. *See* Exotropia
Wandering eye, 173. *See* Amblyopia
Wayne, Harry, 82
Wayne Saccadic Fixator, 82
Weinberger, Norman M., 232, 233
Westheimer, Gerald, 210, 211
Wheatstone, Charles, 4–9, 23, 180
Wick, Bruce, 193, 209, 225,
    226–227
Wiesel, Torsten, xiii, 11, 136–139,
    146–147, 148, 149, 179, 181,
    196, 216, 217, 218, 219,
    221–222, 223, 224
Worth, Claud, 135, 215
Woznysmith, Eric, 77, 123
Wybar, Kenneth, 189, 197

Yale New Haven Hospital, 18, 31

Zack, Richard, 142
Zee, Dvid S., 200
Zeki, Semir, 209